# PRAISE FOR
## *Inner Gardening*

"*Inner Gardening* is a treasure whether you garden or not. If you do, you will be nurtured a hundredfold. If you don't, you will be enticed to plant a paradise of yourself anew."

—Mardeene Burr Mitchell, coauthor of *Hidden Faces of the Soul*

"Diane Dreher delights in the natural world and never forgets that we are part of it, needing challenge, nourishment, pruning, and even times in which we lie fallow. *Inner Gardening* is . . . a lovely, lively book that is as practical as it is inspirational."
—Ron Hansen, author of *Mariette in Ecstasy*

"*Inner Gardening* shows how your earthly gardening labors can be a parallel guide for reaping a harvest of blessings in your inner garden of the soul. A marvelous monthly calendar for growth and renewal in both spiritual adventures."

—Arlene Goetze, M.A., spirituality editor, *Catholic Women's Network*

Robert Numan

## About the Author

DIANE DREHER, PH.D., is the author of *The Tao of Inner Peace, The Tao of Womanhood,* and *The Tao of Personal Leadership.* She holds a Ph.D. in English from UCLA, with credentials in spiritual counseling and holistic health. Diane leads workshops on balance and personal growth nationwide. She teaches Renaissance literature and creative writing at Santa Clara University and cultivates her garden at home in the San Francisco Bay Area.

# INNER GARDENING

*A Seasonal Path*

*to Inner Peace*

*Diane Dreher*

Quill

*An Imprint of HarperCollinsPublishers*

For Robert,
*con amore*

Thy firmness draws my circle just,
   And makes me end where I begun.

       JOHN DONNE,
   "A Valediction Forbidding Mourning"

# Contents

# Acknowledgments

I WOULD LIKE to thank Betty and Harold Johnson for planting the garden I love and my husband, Robert Numan, for helping me cultivate it on many levels. I would also like to thank the many people who have shared their insights about gardening: my mother, Mary Ann Dreher; and my friends Christiaan Lievestro and Ron and Rhonda Schlupp; along with Pat Welsh, author of *Southern California Gardening*; Don Ellis, resident horticulturist at the Elizabeth Gamble Garden Center in Palo Alto, and the Master Gardeners of Santa Clara County. For insights into the relationship between outer and inner gardening, I am indebted to generations of English poets and to Stanley Stewart of the University of California, Riverside; Paul Jorgensen, Geneva Phillips, and George Guffey of UCLA; and Earl Miner of Princeton University, who introduced me to them years ago. I also benefited greatly from research in the gardens and rare book collections of the Henry E. Huntington Library in San Marino, California, and from my friendship with Janette Lewis, when we did our dissertation research there. For advice and information about medieval and Renaissance history, gardens, and spirituality, I am grateful to my friends and colleagues Ann Brady, Phyllis Brown, Rev. Brad Bunnin, Bo Caldwell, Tina Clare, Judy Dunbar, Michael Fernandez, Ron Hansen, Tom Judd, Tracey Kahan, Ed Kleinschmidt Mayes, Catherine Montfort, Elizabeth Moran, Richard Osberg, Theodore Rynes, S. J., Peggy Saad, Sunny Skys, William J. Sullivan, Tom Turley, Katherine Woodall, Cory Wade,

Fred White, and my research assistant, Anna Chadney. For insights about contemplation and spiritual growth, I would especially like to thank Sister Ann Wittman, S.C.S.C., of Minocqua, Wisconsin; and William J. Rewak, S. J., director of El Retiro San Iñigo, the Jesuit retreat house in Los Altos, California.

I would like to thank my agent, Sandy Dijkstra, for her encouragement and belief in this project and my editor at HarperCollins, Toni Sciarra, for her editorial artistry which helped see this book to fruition. Finally, because all of our gardens are part of the larger pattern of life on this planet, some of the royalties from this book will be used to help preserve this beautiful planet we call home.

# INNER
# GARDENING

# ONE

# *Gardens and Personal Growth*

We must cultivate our own gardens.

VOLTAIRE,
*Candide* (1759)

WELCOME TO *Inner Gardening,* a book about growth and cultivation on many levels. Whether you're a new gardener thinking of growing vegetables in your backyard, a more experienced gardener curious about garden history, or just someone who's always loved plants, you will find in these pages practical advice and information along with something every gardener has realized: that cultivating the soil can be a powerful spiritual exercise. Working in our gardens takes us on a journey of discovery within and around us, deepening our connection to nature and ourselves.

Gardening offers a natural remedy for the escalating stress of contemporary life. Pushed along by our noisy, busy culture, too many people equate self-worth with productivity, expecting their bodies and minds to work incessantly, like machines. Yet beneath the high-tech, high-stress surface of our lives, we still move by nature's rhythms, live by nature's cycles. Unlike machines, each of us has our own circadian rhythms, daily energy highs and lows, as well as a very human need for balance, which we ignore at our peril.

The seasons of the year and the subtle energies of the sun and moon affect us. We are one with the universe, composed of the very elements of the stars. The blood in our veins, the planets in our solar system, and the water that sustains life on earth move in cyclical patterns. Our creative endeavors move in cycles too: artists, writers, scientists, and inventors from Coleridge to Stravinsky, Edison to Einstein, have experienced the creative process of preparation, incubation, inspiration, and verification. All of our lives have their seasons, their cycles of growth and renewal—if we can only recognize them.

In the past observing these cycles was much simpler. From the Middle Ages through the Renaissance, human life was governed by the cyclical drama of nature. People lived closer to the land, beginning their days with the dawn, returning home at sunset, charting the course of their lives by the four seasons of growth, harvest, dormancy, and rebirth. We cannot return to those simpler times, but we can follow the path of poets, philosophers, saints, and sages, wise men and women throughout the ages who have found inspiration in gardening.

## GARDENING AS SPIRITUAL PRACTICE

GARDENING HAS HELD deep significance in the spiritual traditions of East and West. Twenty-six centuries ago the Taoist sage Lao-tzu recorded the wisdom of nature in the *Tao Te Ching*. Meditating on the unity of creation, Taoist gardeners discovered the deeper harmonies of nature. Chinese garden design has affirmed their vision, from the early Chin dynasty in 221 B.C.E. to our own time. In Japan many Zen monks have been dedicated gardeners.[1] With their emphasis on order, simplicity, and mindfulness, Zen gardens promote a deep meditative awareness.

In the Western tradition gardening dates back to the beginning of time, in the biblical Garden of Eden. Like Adam and Eve, early Christian hermits were gardeners. In 305 C.E., St. Antony of Egypt founded the first

Christian monastery, establishing a tradition that combined prayer with gardening.[2] In the sixth century the Rule of St. Benedict balanced gardening, prayer, and intellectual work, becoming a model for Western monasticism.

Not only saints but secular men and women have found renewal in their gardens. English monarchs who have loved gardening include Queens Eleanor of Provence, Eleanor of Castile, and Philippa of Hainaut, and King James I. The Italian poet Petrarch was an avid gardener. So was Cosimo de Medici, whose garden at Careggi became a meeting place for Renaissance humanists.[3] In England, Sir Thomas More entertained his friend and fellow humanist Erasmus of Rotterdam in his garden in Chelsea, where he described the rosemary growing up his garden walls as "sacred to remembrance and to friendship."[4]

Gardens offer solace and renewal to the world-weary. In the words of the seventeenth-century English poet Andrew Marvell, "All flowers and all trees do close/ To weave the garlands of repose."[5] In their gardens the burdens of the world have fallen from the shoulders of many busy leaders, including George Washington, who grafted fruit trees at Mount Vernon, and Thomas Jefferson, who kept a journal of his garden at Monticello, recording the annual return of herbs and flowers as he struggled with political turmoil and the loss of his beloved wife. In the twentieth century, as storm clouds gathered for World War II, Winston Churchill found strength and perspective working in his garden at Chartwell, his country home in Kent.

## GARDENING AS PERSONAL RENEWAL

GARDENING SLOWS US down, puts us back in touch with our own natural rhythms, teaches us patience and perseverance. It is an old Buddhist practice to plant a tree and tend it, developing a relationship of kinship, compassion, and respect that links us more deeply to the natural world and to ourselves.

Years ago I learned this lesson when my students Eric and Brendan gave me a Japanese maple for my birthday. I returned home one evening in May to find this five-foot tree in a pot by my doorstep, its delicate green leaves shimmering from a light spring rain. Through many annual cycles I cared for that tree, watered and pruned it, watched its leaves turn flaming red in autumn, then fall to the ground as the tree grew dormant. That first winter, as I found my way through a painful divorce, I moved my tree to a small apartment, where it stood outside my door. When friends came to visit they told me, "Your tree is dead," so I tied a note to its branches: "Dear Friends: Do not let my bare branches fool you. I am only hibernating for the winter and will bloom again in spring." That spring the tree came back to life, its tiny red buds opening into a profusion of bright green leaves. Waving gracefully in the breeze, its delicate branches seemed to greet me when I came home each day.

I brought the tree to my new condominium, where it stood in its pot in a small Japanese garden I planted in the backyard. For years the tree kept watch as I left for work each morning and returned at night. There it stood while an old friendship blossomed into love and the seasons of my life changed once more. The year I became engaged, the tree became potbound, so I planted it in the ground, giving it a permanent home. There it stands, in the garden I left behind for the young woman who bought my condominium.

When I married Bob, I moved into his two-bedroom house in San Jose, putting most of my furniture in storage. Two years later, when we bought our new house in Los Gatos, there was a beautiful Japanese maple growing inside the courtyard by the front door. The tree's delicate leaves fluttered in the soft summer breeze. I smiled in amazement. It was like finding an old friend and coming home.

Caring for a plant and watching it grow shifts our attention from the noisy industrial world to the essential patterns of nature. Gardening literally grounds us, returns us to more natural rhythms, teaches us important lessons, restores our faith in life. It strengthens us both psychologically and physically, helping us live longer, healthier lives.[6] Restoring our connection with nature's cycles, gardening gives us a powerful metaphor for living: we cultivate our inner resources as we cultivate the soil, growing stronger, wiser, and more whole.


*Inner Gardening* was written for the gardener in all of us, combining personal narrative with practical advice, inspiration, and traditional garden wisdom. If you are a new gardener, you will find helpful advice about soils, planning, and plant care. If you're more experienced, you will discover a new kinship with gardening traditions, realizing how many tasks you perform in your garden today reach back through the centuries. You'll also find how often gardening brings up essential lessons in self-cultivation, discovering insights to inspire you as you read, reflect, and cultivate your inner garden.

Gardening is a solitary pursuit, offering us much-needed time for reflection. But sometimes, when we're in the midst of some new gardening task—dividing perennials, transplanting a tree, designing a new garden bed—what we really need is a friend, another gardener, to talk to, question, and learn from. In the spirit of friendship I invite you to share a year in the life of my garden. My Northern California garden was a gift from Betty Johnson, the artist who planted and tended it for forty years until she and her husband retired and moved to a small town on the Sacramento delta.

My garden is a green retreat, sheltered by oak trees, enclosed by a redwood fence and temple gate. I have always loved gardening, but living with this garden has been an ongoing education. I will share many of its lessons in this book. Although our gardens are probably miles apart and in different climatic zones, the year's journey through my garden includes yours, for descriptions of tasks in my Northern California garden are supplemented by advice for other regions of the country. In addition, the book is organized by season, with checklists for spring, summer, autumn, and winter. Regardless of regional differences in degree or duration, the seasonal cycles include us all. So no matter what month "early spring" begins in your garden, this is the season to start preparing the soil and planting.

Setting our gardens within a larger context, Chapter 2 looks at garden history, taking you back to the enclosed gardens of the Middle Ages and the Renaissance, when gardening was a familiar metaphor for self-cultivation and spiritual growth. The next twelve chapters take you through the twelve months of the garden year, describing the annual round of tasks and traditions. The heart of the book is this cyclical pattern of four seasons and twelve chapters, leading from the new beginnings of early spring to the warm profusion of summer, the rich harvest of autumn, the quiet wisdom of winter, and, finally, the promise of another spring.

As everyone realizes who has ever planted a seed and watched it grow, what we cultivate around us, we also cultivate within us. Our gardens provide us with harvests of more than fruits and flowers. They give us moments of quiet joy, beauty, and inspiration, as well as lessons in personal empowerment. Enriched by the wisdom of nature's cycles, our lives take on greater meaning and depth. We become more mindful about how we plant and cultivate the seeds of new endeavors, developing greater patience and perseverance, gleaning a wealth of insight from the gardens of our lives.

Affirming this perennial wisdom, each chapter combines advice on cultivating your garden with principles for cultivating greater joy and agency in your life. Practical sections on "Garden Growth" and "Garden Tasks" are accompanied by "Gardening as Spiritual Practice," which combines personal narrative with insights, advice, and prescriptive exercises you can use in your life today. "Garden Reflection," an inspirational narrative, concludes each chapter. Garden quotations from Medieval and Renaissance poets appear in the borders with references following the chapter notes at the end of the book.

*Inner Gardening* can be read in many ways. You can explore the garden year, following the annual cycle from one springtime to the next by reading the book straight through from this chapter to the end. Or you can key your reading to the current season, reading Chapter 2, then moving to the gardening checklist for the season you're in now, following its advice in your garden. Next, you can turn to the chapter for this month, reflecting on the lessons it holds. If you begin reading this book in July, for example, you'd consult the checklist for "Summer," then turn to Chapter 7, "July," to begin your odyssey through the book. You may prefer to read the book straight through, then reread it, one chapter at a time, beginning with the month you find yourself in now, experiencing both the context of *Inner Gardening* and the garden year as it unfolds for you.

However you read *Inner Gardening*, I invite you to enjoy this journey through a year in the life of our gardens, a journey back to the gardens of the Middle Ages and the Renaissance, and, ultimately, a journey of self-discovery as you explore the gardens within and around you.

# Two

## Gardens Past and Present

How could such sweet and wholesome hours
Be reckoned but with herbs and flowers?

ANDREW MARVELL,
"The Garden," ll. 71–72 (c. 1650)

AS YOU AND I walk through our gardens, feeling the summer breeze caress our skin and rustle through the trees above our heads, as we watch a tiny hummingbird dart among the blossoms, or pick the first ripe tomatoes of the season, we follow a tradition old as civilization itself. Dating back at least seven thousand years, the earliest gardens were grown for food. But the *art* of gardening, of participating in the annual cycle of the seasons and watching things grow, has become a perennial source of peace, inspiration, and personal renewal.

What draws you into the garden? Its beauty and serenity? The promise of fresh fruits and vegetables? Do you cultivate favorite flowers: tulips, dahlias, or roses? Do you enjoy learning about the lives and habits of plants? Is your garden a contemplative retreat where you can leave the noisy world behind, returning to the quiet wisdom of nature?

Gardens have been contemplative retreats since the Middle Ages, when monks began the tradition of inner gardening that inspired this book. As the poet Andrew Marvell described, gardens can take us away from the discord of the world outside, "Annihilating all that's made/ To a green thought in a green shade."[1] Green, the liturgical color of hope, was believed to touch our hearts and awaken our souls. The scholastic philosopher Albertus Magnus wrote in the thirteenth century that nothing refreshes the sight so well as a bright green lawn. Looking out at the green grass in a cloister garden, medieval monks rested their eyes and renewed their spirits after long hours of study.[2] How many refreshing shades of green do you find in your garden today?

> *I look upon the Pleasure*
> *we take in a Garden as*
> *one of the most innocent*
> *Delights in humane life.*
> JOSEPH ADDISON,
> *The Spectator* (1712)

In our gardens the simple tasks of watering, weeding, and watching things grow renew us on many levels. Research has related gardening to an increased sense of agency, vitality, and self-esteem. One study of older women noted for their energy and accomplishment revealed that all of them loved gardening.[3] Working in our gardens brings us closer to nature, relieves stress, and gives us the power of positive reinforcement. Planting seeds and seeing our plants grow and blossom shows that our actions *can* make a difference. Such reinforcement is hard to find in a world where much of our work deals with intangibles: words, images, sales, and services. Unlike our grandparents, few of us feel a sense of craftsmanship: building something with our hands, watching it take shape, and admiring the finished product. Most of us work in offices, factories, service industries, or electronics, continuously busy but rarely experiencing agency. Yet after a day dashing from one intangible challenge to the next, coming home to discover a row of tiny seedlings, to water or tend our gardens assures us that, if only in this small plot of ground, we can make a positive difference in our world.

For many of us gardening has become a spiritual practice, a way of cultivating greater peace and order. Even the simple act of weeding can become a contemplative ritual. Years ago, when I began graduate school at UCLA, I moved to a small apartment in Santa Monica with a lovely garden where the landlord's mother grew roses, orchids, and colorful annuals. One day when I returned from a busy day of classes, I asked her if I could pull weeds in the garden. Smiling, she told me, "You're welcome to pull as many as you like." So after my classes in Anglo-Saxon, bibliography, and Renaissance literature, I'd come home and pull weeds. This simple ritual helped me relax, opening my mind to new ideas. That year I did better than I'd expected, getting A's in all my classes. Somehow, clearing away the weeds and invasive grasses convinced me I could handle new academic challenges as well.

## PARADISE AND THE WESTERN GARDEN TRADITION

"TIME BEGAN IN a garden," says the small wooden sign by my garden gate. In the Judeo-Christian tradition life began in the Garden of Eden. Centuries of European poets and artists have linked gardens with this biblical archetype, the earthly paradise where humans lived in harmony with all creation and life was perpetual summer, without sin, struggle, or sickness. Here people remained forever young, with no sadness or pain, and even the roses had no thorns.

The word *paradise* once meant "an enclosed garden." Originating with the Persian word *pairidaeza* (enclosed park), it became the old Hebrew *pardes* and Greek *paradeisos*, which referred to both the Garden of Eden and Heaven itself. Medieval European churches and monasteries had their paradises: enclosed gardens, which grew roses and lilies for the altar and served as quiet meditation retreats.[4]

Medieval monasteries developed the tradition of formal gardens we still enjoy today. Each monastery had its chapel paradise as well as orchards, vineyards, kitchen and medicinal gardens, and the central courtyard or cloister garth. Surrounded by the four covered cloister walks, planted with grass and flowers, and often graced with a fountain or devotional statue, the

*A Garden was the Habitation of our first Parents before the Fall. It is naturally apt to fill the Mind with Calmness and Tranquillity, and to lay all its turbulent Passions at Rest. It gives us a great Insight into the Contrivance and Wisdom of Providence, and suggests innumberable Subjects for Meditation.*

JOSEPH ADDISON,
*The Spectator* (1712)

cloister garth was the center of monastic community life.

Like many of our gardens today, the medieval monastery garden was a *hortus conclusus,* enclosed by walls, hedges, or fences, which kept the secular world outside and protected the plants from vandals, pests, and strong winds. Medieval gardeners loved their boundaries. Their farms were bordered by hedges and ditches; their orchards, vegetable and herb gardens were enclosed by basketlike wattle fences, woven from willow or holly boughs.[5]

*As a garden is my Mind enclosed fast.*
    ANNE COLLINS,
"Song," from *Divine Songs and Meditacions*
(1653)

Medieval religious orders developed the practical and spiritual traditions of gardening. The Benedictine order, founded in 530 by St. Benedict, modeled their monasteries after Roman villas, self-sustaining agricultural communities, with vineyards and gardens for fruits, vegetables, herbs, and grain. St. Bernard, a member of the Cistercians, a branch of the Benedictines founded in the twelfth century, taught that gardening could become a spiritual exercise.[6]

Gardens have long been sacred spaces for prayer and meditation. St. Francis de Sales, St. John of the Cross, St. Teresa of Avila, and St. Catherine of Siena practiced solitary meditations in gardens. In devotional art and poetry from the Middle Ages through the Renaissance, enclosed gardens took on powerful symbolism, representing the biblical Song of Songs and the soul in communion with God. The thirteenth-century Franciscan St. Bonaventura taught people to meditate on the natural world. In the sixteenth century St. Ignatius Loyola, founder of the Jesuits, ended his *Spiritual Exercises* with a vision of all creation permeated by the loving presence of God.[7]

*Like to a garden that is closed round,*
*That heart is safely kept.*
    CHRISTOPHER HARVEY
*The School of the Heart*
(1664)

Today some gardens still reflect these early spiritual traditions. The twenty-one California missions, extending from San Diego to Sonoma, were established by Spanish Franciscans over two centuries ago. Every spring at Santa Clara University, founded by the Jesuits at Mission Santa Clara, the wisteria blooms along the old mission walks, filling the pergola-sheltered pathways with clusters of fragrant blossoms. Some venerable vines have trunks larger around than my body and appear to be centuries old. Roses bloom along the garden paths, bordered by colorful pansies and impatiens. When I began teaching at Santa Clara in the 1970s, I'd often dash across campus to an early class and see an old priest quietly walking with his breviary, saying the Divine Office in the early morning mist.

On the hillside above my home in Los Gatos is the old novitiate where novices once lived, studied, and worked in the vineyards of the Jesuit winery. Years ago at daily Mass when my Jesuit colleagues said together over the sacramental wine "Fruit of the vine and made with human hands," the words were quite personally true, for many had spent their early years tending the novitiate vineyards.

The novitiate winery is closed now, many of those Jesuits are gone, and new generations of college students bring their skateboards and cell phones into the mission gardens. But on some quiet mornings the contemplative spirit still remains.


## MEDIEVAL PLEASURE GARDENS

THROUGHOUT OUR HISTORY, as times have changed, gardens have served more secular purposes. By the thirteenth century the European nobility had their own enclosed gardens, modeled after the medieval cloister garths. Medieval ladies grew herbs and vegetables in small raised beds and a variety of flowers in their *herbers*, or pleasure gardens.[8]

As we can tell from their tapestries, paintings, and stained-glass windows, medieval men and women loved bright primary and secondary colors and were familiar with all kinds of flowers. The artist Candace Bahouth has noted in medieval tapestries over a hundred varieties of flowers, so accurately portrayed that botanists can still identify them.[9] Roses, daisies, daffodils, foxgloves, columbine, hollyhocks, lavender, irises, wild strawberries, lilies of the valley, madonna lilies, periwinkles, violets, and primroses were favorite medieval flowers.

Medieval ladies prized flowers for their fragrance as well as their beauty, wearing floral garlands and sprigs of rosemary and lavender. They scattered aromatic herbs on the floors of their homes to sweeten the air. King Henry VI's court physician recommended that houses be scented with roses, violets, mint, and bay, believing this would help prevent disease. At least it made the air more pleasant in a time without modern plumbing or personal hygiene. To repel fleas, medieval men and women scattered pennyroyal (mint) around the house or wore it in their clothes. They brushed their teeth with rosemary paste and ran rosemary branches through their hair as combs. Ladies dried flowers for sachets and made flower perfumes, medicines, and cosmetics. Flowers were also part of the medieval diet, which included rose and lavender syrups, candies, jellies, puddings, and preserves. People ate lavender flowers in salads and cooked them in poultry stews; they used pot marigolds for medicine and cooked their yellow blossoms in stews and pottages.[10]

In millefleur tapestries that captured the beauty of their gardens, medieval ladies stitched bright blossoms of red, blue, and gold, depicting flowery meads, the mixture of grass and flowers growing in their courtyards. In the winter these tapestries brightened castle walls and helped insulate dank rooms from damp and cold, simulating a little Eden indoors. In the spring and summer pleasure gardens became popular outdoor chambers where lords and ladies would socialize, relax after dinner, or share a light meal. People sat on the ground on cushions or on garden benches covered with sod, surrounded by sweet-smelling herbs and flowers.[11] Filled with roses, violets, honeysuckle, and other strongly scented flowers, as well as with lavender, rosemary, and pennyroyal, these enclosed gardens were highly aromatic. Planted with flowers and herbs, the lawns were fragrant outdoor tapestries to walk, sit, or lie upon.[12]

By the late Middle Ages pleasure gardens had become romantic retreats for knights and ladies who practiced the art of courtly love. According to the philosophers, the lady was supposed to inspire her knight to perform virtuous deeds, but many gardens became sites for amorous assignations. Henry II of England met his mistress Rosamund in the elaborate labyrinth of his garden at Woodstock until discovered by his queen, Eleanor of Aquitaine, who had the bower destroyed.[13] A letter still exists from Henry VIII asking Anne Boleyn to meet him secretly in the garden. After his break from the Catholic Church, Henry ordered the dissolution of English monasteries. As he distributed their lands among his favorites from 1536 to 1540, the old medieval cloister gardens became the pleasure gardens of the new Renaissance aristocracy.[14]

> May I a small house and
> large garden have!
> And a few friends, and
> many books, both true,
> Both wise, and both
> delightful too!
> ABRAHAM COWLEY,
> "The Wish," ll. 10–12
> (1668)

During the Renaissance pleasure gardens were enjoyed throughout Europe, not only by the aristocracy but also by the growing middle classes. Retaining their enclosed rectangular design, Renaissance English gardens were decorated with box hedges clipped in elaborate knot designs, with topiary and mazes as well as classically inspired statues and fountains. World explorations brought in colorful new flowers from America and the Middle East to brighten Renaissance gardens. In 1580 hyacinths arrived in England from Turkey via Padua, and tulips, also from Turkey, came in from Holland. By 1597 sunflowers had been brought over from North America and nasturtiums from South America. The Renaissance also inspired a new scientific curiosity about plants. The first European botanical gardens were established at the University of Padua in 1545, with others following on the Continent, and the first English botanical garden was established at Oxford in 1621.[15]

## SEVENTEENTH-CENTURY DEVOTIONAL GARDENS: THE PARADISE WITHIN

IN CHALLENGING TIMES people turned to their gardens for peace and consolation. After their defeat by the Puritans, English Royalists embraced a simpler life and stoic philosophy on their country estates. John Dryden later echoed these sentiments in his translation of Virgil's *Georgics*, which contrasted the noise, politics, and corruption of city life with

> easie Quiet, a secure Retreat,
> A harmless life that knows not how to cheat,
> With homebred Plenty the rich Owner bless,
> And rural Pleasures crown his Happiness.
> Unvex'd with Quarrels, undisturb'd with Noise,
> The Country King his peaceful Realm enjoys.[16]

Many Puritans, from William Prynne to John Milton, whose daily routine included long walks in the garden, also found inspiration in their gardens. Milton had a garden growing wherever he lived. The elaborate floral descriptions in his poems reveal his profound love of gardens.

Near the end of Milton's *Paradise Lost*, Adam and Eve learn that they must leave the earthly paradise but will find a "paradise within," in the cultivation of their souls.[17] This paradise within was sought by many people who turned to their gardens for inspiration through the seventeenth and eighteenth centuries. With its seasons of renewal and daily lessons in patience and hope, the garden was seen as a living guide to devotion.

*How well the skillful gardener drew,*
*Of flowers and herbs, this dial-new;*
*Where, from above, the milder sun*
*Does through the fragrant zodiac run;*
*And, as it works, the industrious bee*
*Computes its time as well as we!*
*How could such sweet and wholesome hours*
*Be reckoned but with herbs and flowers?*
ANDREW MARVELL,
"The Garden," ll. 65–72
(c. 1650)

Garden furnishings, arbors, and paths developed symbolic meanings. Tracing the course of the sun and, by implication, the course of human life, the garden sundial reminded people to look beyond the superficial and make wiser use of their time. Herb gardens were often planted in the shape of a large sundial.

## GARDENING AS SELF-CULTIVATION

IN THE SEVENTEENTH and eighteenth centuries gardening was seen as a means of building character. Milton's "Of Education," written in 1644, combines a traditional classical education with practical lessons in gardening. He recommended that people read "the authors of agriculture," Cato, Varro, Columella, and Virgil (*Georgics*), who would give them "natural knowledge ... they shall never forget, but daily augment with delight."[18]

*Happy the Man, who, studying Nature's Laws, Thro' known Effects can trace the secret Cause.*

JOHN DRYDEN, *Virgil's Georgics* II, 11698–701 (1697)

In the next century Alexander Pope equated gardening with self-discipline, becoming one of the major gardening authorities of his time. He wrote that "my garden, like my life, seems to me every year to want correction and require alteration. I hope, at least, for the better." Pope moved into his house in Twickenham on the Thames in 1719, when he was thirty-one and spent the rest of his life there writing and cultivating his garden. He made a lifetime study of garden design and raised vegetables and pineapples, which he shared with his friends.[19]

The eighteenth century opened up the once enclosed English garden, providing a new scope in garden design that swept across the landscapes of large country estates like Stowe and Blenheim. Instead of fences and walls garden designers like Capability Brown used the sunken ditch or ha-ha to divide property, conveying a new sense of grandeur and space. Like the later political revolutions in America and France, this English gardening revolution affirmed new principles of natural order, reason, and liberty.

In America, Thomas Jefferson studied botany as a young man and kept careful records of his garden at Monticello. He took a tour of English gardens with James Madison and studied continental gardens while serving as ambassador to France, bringing many European plants back to his own garden. Jefferson wrote in later years that "no occupation is so delightful to me as the culture of the earth, and no culture comparable to the garden. I am still devoted to the garden. But though an old man, I am but a young gardener."[20]

Jefferson knew then what you and I have since realized: that living with a garden is an ongoing education, filled with insights that keep us forever in discovery, and in this sense forever young.

## Creating Your Own Garden Tradition

GARDENS HAVE SERVED many purposes over time, providing food, flowers, an intimate relationship with nature, and a deeper knowledge of ourselves.

### *Personal Exercise*

AFTER THIS BRIEF tour of garden history, take a few minutes to consider which garden traditions you'd like to cultivate for yourself. In your garden journal or on a piece of paper, answer these questions:

- What aspect of garden history appeals to you most?
- How can you apply this tradition to your own garden?
- Would you like
  - a "show garden," with attractive foundation planting and a border of flowers in your front yard?
  - a kitchen garden for fresh herbs and vegetables?
  - a paradise or quiet place for contemplation?
  - a learning laboratory to grow new and exotic plants?
  - a cutting garden to provide the beauty of cut flowers in the house
  - a pleasure garden, a pleasant outdoor space to entertain family and friends?

Keep these notes to refer to as we explore the garden year. In each season you will discover new ways to cultivate the garden of your desire.

## GARDENING AS THE ART OF THE POSSIBLE

KNOWING WHAT YOU want in a garden is the first step toward creating your personal paradise. Next you must consider what is actually possible. What do you have to work with?

I've never had the capability or the inclination for an expansive eighteenth-century garden. When I moved into my condominium in the early 1990s, I had only a six-by-nine-foot raised bed along the redwood deck in my backyard. There I discovered the joy of working in miniature, creating a small meditation garden. Setting my potted Japanese maple near the back fence, I selected some interesting rocks from the local rockery, carried them home in the trunk of my car, and arranged them around the tree. I pruned one azalea bush like a bonsai tree, removed some unruly shrubs, added two smaller azaleas, and surrounded the plants with a "stream" of white gravel. A stone lantern completed the design. Since I also wanted a kitchen garden, I planted tomatoes and strawberries in pots along the other side of the deck, then hung some potted geraniums along the fence. My miniature garden was a daily delight, providing me with color, shade, a cool green corner for contemplation, and fresh tomatoes every summer.

As gardeners we are artists who work with nature. Sometimes you'll have a blank canvas; sometimes you'll inherit someone else's design. This summer our friend Jeff moved into a brand-new house in Benecia, where he has a backyard full of potential. Right now the yard has only a wooden perimeter fence, lots of bare earth from recent construction, a few wildflowers, and a grill. But Jeff is making his garden plan and can already visualize a stand of trees—dark green cypress and Japanese maples— planted next to the fence, forming his own small forest and sheltering his yard from the strong afternoon winds.

When Bob and I moved into our house in Los Gatos, our paradise was provided by the previous owners, Betty and Harold Johnson. Inside the front courtyard is a beautiful Japanese maple, along with a variety of ferns, azaleas, and heavenly bamboo. Ajuga covers the ground with a tapestry of green tinged with plum. Pink and white azaleas bloom in the spring, followed by the sunny yellow blossoms of St. John's wort.

Around the north side of the house are more azaleas, ferns, and an avocado tree Betty planted from a seed. Here daffodils bloom in early spring, followed by pink and red camellia blossoms on the bushes beside the house. Growing near the back garden gate is enough mint for many summers of iced tea.

Our backyard, encircled by a greenbelt and large oak trees, feels like a private retreat. A redwood deck fills most of the yard, with blue spruce, liquidambar, and wisteria providing shade on the northern side. The deck is bordered in back by Norfolk Island pines, a fragrant bay tree, and many flowering shrubs. A mandarin lime tree, azaleas, and rhododendrons grow in the garden on the south side of the deck, and a small lemon tree stands in an enclosure near the stairs. In the southeast corner stands a tall redwood tree that Betty planted as a seedling. The south side of the house is bordered by more liquidambar, Japanese maples, ferns, and azaleas.

The front yard blooms with colorful flowers that change with the seasons. Spring offers a parade of bulbs and perennials: crocuses, grape hyacinths, freesia, hyacinths, anemones, and tulips on both sides of the cobblestone path to our front gate. Later irises bloom beneath the birch trees on the northern side. Along the sidewalk and the street, over two dozen rosebushes bloom from May through October. Annual California poppies and lacy white alyssum reseed themselves, filling in when the bulbs have died down, and a team of dahlias joins the roses in late summer.

To the south, between our driveway and the yard next door, is a friendship garden planted by Betty and our neighbor Rhonda Schlupp. This garden blossoms with poppies and perennials from early spring until late autumn. In midsummer it is ablaze with orange crocosmia and blue-violet lilies of the Nile.

Our yard combines many garden traditions. The front gardens welcome friends with roses and other floral displays, providing cut flowers to enjoy inside. We can step outside to the contemplative serenity of a "green shade" within the quiet front courtyard or around our redwood deck in back. The deck can also become a pleasure garden for barbecuing, outdoor dining, and entertaining friends. The only thing missing was a kitchen garden, so I planted herbs among the annuals in the front yard, tomatoes along the sunny south side yard, and green beans near the north garden gate.

When I first came here I was filled with a deep sense of responsibility for all the gardens. It was a hot September when we moved in, and I ran around the yard every day with the garden hose and sprinkler, trying to keep up with all the thirsty flowers. I had grown roses, annuals, and vegetables for years but had never had such a profusion of perennials, hidden secrets that magically emerged from the ground, their bright blossoms often catching me by surprise. That first year I became a botanical detective, investing in books on plant species. I wanted to get to know all these new neighbors so I could take proper care of them. I ended up learning more than I'd anticipated, not only about my garden but about myself. Seeing my own life in terms of cultivation, sowing, pruning, and seasonal cycles, my eyes have opened more fully to the world around me and the perennial wisdom of nature.

In this spirit of continuing growth and discovery, I encourage you to experiment with your own garden and invite you to join me on a journey through the garden year as we explore the gardens within and around us, creating our own personal garden traditions.

# SPRING

# Spring Garden Checklist

## Early Spring

*Plant:* Plant bare-root roses, trees, and shrubs. Buy seeds and bulbs. When weather warms up, sow sweet peas and hardy annuals; set out hardy perennials. Plant onion sets and seed potatoes. Set out strawberry, broccoli, and cabbage plants. Sow cool-weather vegetables (check seed packets for correct planting time for your area). Use cold frames in colder areas. Start seeds for warm-weather flowers and vegetables indoors. Plant container-grown shrubs and trees when danger of frost has passed.

*Water:* With spring rains you may not have to water as much, but give special attention to seedlings and transplants.

*Weed:* Pull weeds while their roots are still small and the soil is moist.

*Feed:* Feed established fruit bushes, shrubs, and trees before spring growth begins. Feed roses, early bulbs, and flowers in beds and borders as they begin to grow.

*Cut back:* Prune roses if still dormant. Thin vegetable seedlings. Inspect frost-damaged trees and shrubs for new growth, then prune away deadwood. Prune spring-flowering shrubs after they bloom.

*Cultivate:* Dig new flower beds and cultivate the soil, adding compost and other nutrients. If summer and fall-blooming perennials have become overcrowded, divide and replant them. If nights are still cold, cover seedlings and tender plants. Mulch fruit bushes with compost. Remove winter protection from trees and shrubs as weather warms up.

*Harvest:* Harvest the first green onions from your garden.

*Watch for:* Check for snails and slugs. Set out traps or bait as needed. Pick up fallen blossoms from flowering shrubs to prevent disease.

*Enjoy:* Enjoy dozens of small surprises as your garden comes back to life.

## Mid-Spring

*Plant:* Sow sweet peas, hardy annuals, onions, and potatoes. In cooler regions sow tender annuals indoors. Harden them off before planting outside. When weather warms up, sow or set out frost-tender flowers as well as strawberries, warm-weather vegetables, and herbs (check seed packets for correct planting time for your area). Plant container-grown trees and shrubs, perennials, and summer bulbs.

✂ *Water:* With spring rains you may not have to water as much, but give special attention to seedlings and transplants.

✂ *Weed:* Pull weeds now, when they are still small and the soil is moist. A layer of mulch will help keep down weeds.

✂ *Feed:* Feed roses, ground covers, flowers, and vegetables as they begin to grow. Fertilize spring bulbs when they begin to grow and again after they bloom. Feed fruit trees, and give acid food to azaleas, rhododendrons, and camellias after they've stopped blooming.

✂ *Cut back:* Trim spent blossoms from daffodils and other early bulbs but leave foliage in place. Thin vegetable seedlings as needed. Continue to prune frost-damaged trees and shrubs. Trim ivy and box hedges.

✂ *Cultivate:* Stake tall plants and tie up climbing plants to protect from spring winds. Cultivate the soil, top-dressing with compost. If nights are still cold, cover seedlings and tender plants.

✂ *Harvest:* Some early greens may be ready to harvest. Use thinnings from lettuce, spinach, and green onions in spring salads.

✂ *Watch for:* Check for snails and slugs. Set out traps or bait as needed. Check roses and tomatoes for aphids and spray with insecticidal soap as needed. Pick up fallen blossoms from flowering shrubs to prevent disease.

✂ *Enjoy:* Enjoy the colors of spring and the first salad greens of the season.

## LATE SPRING

✂ *Plant:* Sow or set out annuals to fill in where bulbs have died back. Sow herbs and warm-weather vegetables (check seed packets for correct planting time for your area). Harden off plants grown indoors before planting outside. Plant container-grown trees, shrubs, and perennials. Finish planting summer bulbs.

✂ *Water:* As spring rains decrease and the weather warms up, begin watering more often. Pay special attention to seedlings and transplants. Soak the roots of roses and trees once a week.

✂ *Weed:* Keep weeding regularly.

❧ *Feed:* Fertilize roses, annuals, and perennials. Feed spring bulbs when they begin to grow and after they bloom. Feed summer bulbs when first shoots appear. Continue to feed vegetables regularly. Side-dress with compost. Feed fruit trees, and give acid food to azaleas, rhododendrons, and camellias after they've stopped blooming.

❧ *Cut back:* Shear spent blossoms from annuals to keep them blooming. Trim spent blossoms from spring bulbs but leave foliage in place. Thin vegetable seedlings. Pinch suckers off tomatoes. Pick off spent blossoms, and prune spring-flowering shrubs after they bloom.

❧ *Cultivate:* Stake tall plants. Put straw around strawberries and mulch around plants to conserve water and insulate the soil. Put compost or aged manure around the roots of fruit trees. Mulch trees and shrubs as the weather grows warmer.

❧ *Harvest:* Some early vegetables should be ready to harvest.

❧ *Watch for:* Check for snails and slugs. Set out traps or bait as needed. Check roses and tomatoes for aphids and spray with insecticidal soap as needed. Pick up blossoms from flowering shrubs.

❧ *Enjoy:* Enjoy spring bulbs in bloom and new discoveries in your garden.

# THREE

## March: Cultivating the Gardens Within and Around Us

Daffodils,
That come before the swallow dares, and take
The winds of March with beauty.

WILLIAM SHAKESPEARE,
*The Winter's Tale*, IV.4, ll. 118–20 (c. 1610)

IN MARCH, IN this first springtime in our new home, the front yard is alive with daffodils, waving their yellow heads in the breeze. Every day brings more daffodils: charming miniatures; blooms in butter yellow, white, and gold; poet's daffodils with white petals and yellow centers; white double daffodils, and large-trumpeted golden King Alfreds. Popping up all over the yard, they take me by surprise, flying their glad colors, proclaiming spring.

Daffodils grew in ancient Roman gardens and throughout medieval England, and they were claimed by the Welsh as their national flower, worn in March for St. David's Day. Emerging after long winters, these sunny blossoms have captured the imaginations of English poets from Shakespeare and Spenser to Herrick, Milton, Wordsworth, and Tennyson.

In addition to St. David's Day on March 1, this month's holidays include March 17, St. Patrick's Day, and March 25, the Feast of the

Annunciation, with its promise of eternal life heralded by the blossoms of early spring. Shrovetide generally occurs in March, marked by the wild feasting of Shrove Tuesday or Mardi Gras (literally, "Fat Tuesday"), the day before Ash Wednesday, and the forty-day fast period of Lent. In old England, Shrove Tuesday was a time of carnival (from *carnevale*: a farewell to the flesh), celebrated with feasts of pancakes, games, and drinking. The origin of this feast probably hearkens back to pre-Christian fertility celebrations, held around March 20 to mark the vernal equinox and the beginning of spring.[1]

## GARDEN GROWTH

EARLIER THIS MONTH the daffodils in my front yard were joined by legions of grape hyacinths, raising their tiny purple standards. By late March the garden has become a living tapestry of flowers: red and yellow freesia; scarlet, violet, and white anemones; and a procession of tulips in shades of deep crimson, rose, white, vermilion, butter yellow, tangerine, and lilac, with some blossoms candy-striped in red and gold. An artist with nature's palette, Betty Johnson planted this garden. It was so beautiful when we moved in last fall that people in the neighborhood came by to congratulate us. This month I was so touched by all the beauty I wrote her a thank-you note.

*When Winter's rage*
*abates, when chearful*
*Hours*
*Awake the Spring, and*
*Spring awakes the*
*Flow'rs.*

JOHN DRYDEN,
*Virgil's Georgics*,
I, ll. 463–64 (1697)

Although spring comes in different ways to different regions, its advent is unmistakable. What signs of life do you see in your garden? Plum trees in bloom? Early crocus blossoms peeking through the snow? Paper-white narcissus and snowdrops? New green shoots on trees and shrubs? Look closely, for something is happening, if only beneath the surface. The natural world has begun to awaken again. In Massachusetts the ice is thawing on Walden Pond. In Vermont the sap has risen in the sugar maples.

Spring is a time of new beginnings. The days are brighter, with hours of sun returning after long winter nights. Camellias and azaleas are blooming within my garden walls. There are bright pink blossoms on the dwarf peach tree. Up the hill from my house, other fruit trees have turned from winter bareness into visions of pink, plum, and white. All around us the world is returning to life. What better time to begin our garden year together than in this season of renewal?

## GARDENING AS SPIRITUAL PRACTICE

### *Cultivating Your Inner Garden*

IN THEIR WORLD of poetic symbolism, medieval and Renaissance gardeners found parallels between outer and inner gardening. They saw cultivating the soil as an exercise in patience and humility.[2] Sowing seeds for a new harvest and tending their gardens were acts of faith. Their vision of inner gardening offers a valuable message to us today: our lives make sense when we stay on course, when our everyday actions reflect our deepest values.

What blows us off course? Not always great events and circumstances. As we rush from one commitment to another, we can betray ourselves in hundreds of small ways: surrendering our values, priorities, and boundaries out of overwork, expediency, distraction, or exhaustion. In mindless moments we often make foolish choices. Our lives become cluttered as we reap the results of our indiscriminate sowing.

The alternative is mindful self-cultivation, which returns us to the wisdom of earlier times, when daily life was grounded in spiritual exercise. With their symbolism of herbs and flowers, centuries of poets and philosophers have portrayed the garden of the soul, enjoining people to cultivate their inner lives through mindfulness and regular spiritual practice.

The Renaissance and medieval worlds were informed by the concept of correspondences, relating the individual, or *microcosm*, to the *macrocosm*, or larger world around us. They believed that what we cultivate around us we also cultivate within. Recent research in neuroscience reveals this concept to be much more than metaphor. Our brains develop throughout our lives, growing new neural pathways, perhaps even new neurons, in response to stimulation.[3] With our daily actions and attitudes we literally cultivate ourselves, modifying the physiology of our brains. Like tiny seedlings, with axons and dendrites branching from their seedlike bodies, our brain cells grow and develop, making new connections as we cultivate new possibilities.

This explains why when we first learn something, we are awkward, slow, and uncertain but with practice become more proficient. Do you remember the challenge of learning to drive—managing that mechanical beast with its roaring engine and all its gears, pedals, and gauges? But now we drive fluidly, without even thinking of the separate actions that once confounded us. Why is this? Because with practice we grew new connections, new pathways in our brains. We cultivated that part of our inner garden that allowed us to perform the task. So it is with everything we learn. With practice, we continually cultivate ourselves.

Our daily actions and choices determine who we are and what we become. Learning a new skill promotes greater insight, patience, and precision, while a more passive pursuit like watching television only reinforces passivity. As our medieval and Renaissance ancestors realized, the discipline, challenges, and discoveries of gardening can create new life within and around us.

## The Virtues of Compost

WORKING IN OUR gardens, we discover enduring principles of growth and renewal. A compost pile turns leaves, grass clippings, and kitchen waste into rich new soil. The same principle holds true for our inner lives.

Composting is part of the natural cycle in which nothing is wasted. Apple parings, ends of vegetables, onion skins, tips of green beans, last week's leftovers, vegetables that have gone bad in the refrigerator—all go into the compost bin. No apologies. Whatever they are, wherever they come from, they're part of a larger process of transformation, turning remnants of the past into better tomorrows.

In an examined life everything is compost. Cherished memories empower us and enrich our lives. But so can our mistakes, old habits we'd like to break, patterns we've outgrown. Instead of dwelling on a negative experience, we can compost it. Becoming more mindful, asking, "What can I learn from this?" and then moving on can turn any negative experience into a new cycle of wisdom and growth.

## Personal Exercise: Composting the Past

- This week think of something in your life you'd like to compost:
  - an old habit you'd like to break
  - a negative experience that keeps nagging at you
  - something you did that you regret
- Write it down on an index card: "I compost _____ (name the action)." Then sign and date the card.
- For the rest of the month, look at the card each morning and say to yourself, "I compost _____," stating what you've chosen to compost.
- It takes time to break old habits, so don't be discouraged. Whenever you find yourself falling back into the old pattern, stop and tell yourself, "I've composted that."
- At the end of the month, take the card, tear it up, and bury the pieces in the ground, adding your compost to the soil.

In the process of composting the past circles back to enrich the present. You can become stronger and more confident by composting past mistakes and imperfections. Composting expands our perspective, putting us in touch with the natural cycles that renew our faith in life.

## Weeding Your Inner Garden

IF YOU DON'T keep weeds under control, they'll crowd out all the good plants. The same principle holds true for us. Our lives, like our gardens, have their own structure and design. Weeding out some activities to make room for others is an important spiritual exercise, an ongoing process of personal growth.

Over time we all collect weeds in our lives' gardens: mindless habits, unwelcome intruders, or unproductive activities that drain our energy, interfering with our chosen patterns of growth.

## Questions to Consider

As WE BEGIN a new springtime of growth, March is a good time to identify and root out the "weeds" in our lives. Take some time this week to consider the following:

> It is essential for true discipleship to free oneself from useless cares in the affairs of life. Anyone who is anxious about useless things cannot give attention to those which are profitable.
> ST. BONAVENTURA, "Sermon on St. Francis" (1255)

- Do you have any habits or activities that steal your time and drain your energy? It's time to root them out.
- Ask yourself how you can say no to these things in order to say yes to the rest of your life. Recently, when I hadn't had time to exercise because I'd been leaving work late, I realized why: an unhappy co-worker had been dropping by regularly to complain about a chronic problem. The next afternoon when this person came by, I said I had a deadline and asked if we could schedule another time to deal with the issue. For the first time in weeks, I left work on time. As I headed out to exercise, I knew that being interrupted by gossip was a habit I needed to break.
- What is one "weed" you'd like to eliminate from your life? Figure out a way to root it out. Take action. Clear the ground. Give yourself room to grow.

## Planting Seeds of Renewal

ONCE WE'VE CLEARED the ground we can sow our personal seeds of renewal. Do you have a dream in your heart that has always seemed too big or too far away? Today, embrace that dream and plant the seed.

- Write down your goal in your journal or on a sheet of paper.
- Now think of the steps necessary to get you there. Write them down. Tomorrow you can look at the list again, choosing one small step to take first and writing it on your calendar. Taking one step at a time, you *will* reach your goal and achieve your harvest.
- But for now congratulate yourself. Planting seeds is always the first step.

## GARDEN TASKS

### Cultivating the Soil

FROM MEDIEVAL TIMES to today March has been a time to cultivate the soil. A thirteenth-century Book of Hours pictures people digging in their garden beds and working in the orchard, putting new soil around fruit trees.

Today's gardeners still rejoice in early spring as they return to their gardens after long winter dormancy to prepare the soil for spring planting. The actual time you do this depends on the weather in your region. In milder regions some gardeners are already setting out their first spring flowers and vegetables. Others are waiting, their soil still too frozen or too waterlogged to cultivate. But gradually, as the days grow warmer and daytime temperatures rise above 42 degrees Fahrenheit, spring returns to our gardens. In the Northeast

*While yet the Spring is young, while Earth unbinds
Her frozen Bosom to the Western Winds;
While Mountain Snows dissolve against the Sun,
And Streams, yet new, from Precipices run:
Ev'n in this early Dawning of the Year,
Produce the Plough, and yoke the sturdy Steer.*

JOHN DRYDEN,
*Virgil's Georgics,*
I, ll. 64–69 (1697)

gardeners will soon be removing the winter mulch from their perennial beds and planting their first early peas.

Cultivating our gardens improves the texture of the soil. You can test your soil's texture by picking up a handful and squeezing it. Sandy soil has a light, loose texture. It seeps through your fingers and feels gritty when rubbed. Clay soil, like mine, is sticky, smooth, and dense. Easily molded when wet, it forms a solid mass like the clay you played with as a child. Sandy soil is softer and drains more quickly but contains fewer nutrients, while clay soil has more nutrients and retains water longer—sometimes too long. Clay can become waterlogged when wet and, when dry, too heavy and hard for a plant's roots to penetrate. Although some plants can live in either extreme, the best garden soil is rich loam: a mixture of clay, sand, and silt soft enough for a plant's roots to grow in, yet dense enough to retain water and valuable nutrients.

*Well must the ground be dig'd, and better dress'd, New soil to make, and meliorate the rest.*

JOHN DRYDEN,
*Virgil's Georgics,*
II, ll. 87–88 (1697)

*If earth be not soft, Go dig it aloft.*

THOMAS TUSSER,
*Five Hundred Points of Good Husbandry*
(1580)

I've been working to improve my soil's texture by adding compost. March is a perfect time to do this, when we're out there digging in our gardens. First, pull all the weeds. Then spread a three- to four-inch layer of compost over the soil and dig it in to the depth of your shovel. If you already have bulbs and perennials growing in your garden, you can simply spread some compost around. Your plants will thank you for it. Just be careful not to disturb any new growth.

## Double-Digging

DOUBLE-DIGGING THE soil provides a rich, fertile bed for your plants. Classic French intensive raised beds are about three feet wide, to allow for gardening without trampling down the soil, and double-dug two feet deep. To double-dig your garden bed, you'll need a flat spade, lots of time, and patience initially. But even if you don't manage to dig down twenty-four inches, your efforts will improve the soil dramatically.

## How to Do It

- If you're working in heavy clay soil, first soak the ground so the clay is workable. Remove the weeds and take a sample of soil if you're going to test it. Then spread a one-inch layer of compost across the garden bed.
- Next (or on the next day) dig a trench one foot wide and one foot deep at one end of your garden bed, piling the soil in back. (This is the first dig.)
- Now push your spade into the bottom of the trench, moving it around to loosen the subsoil, ideally another twelve inches but, realistically, as much as you can. (This is the double dig.)
- Repeat this procedure, digging another one-foot trench beside the first, moving the soil from this trench into the first one. Loosen the soil at the bottom of the second trench as you did the first.
- Repeat this process for the rest of your garden bed. Fill the last trench with soil from the first one you dug. Any remaining soil can be used in your compost pile or other parts of the garden.
- Add organic nutrients (such as manure, alfalfa meal, and wood ash) as needed to the top two or three inches of soil and spade them in.[4]
- Admire your work: a beautiful, rich, light-textured soil for your plants.[5]

## Garden Tools

LOOKING DOWN AT my spade as I turn over the soil, I realize that garden tools have changed very little over the centuries. Medieval gardeners had their spades, rakes, and hoes, not much different from the tools you

### Some Basic Garden Tools

- spade (*for digging, cultivating, double-digging*)
- shovel (*for lifting and digging*)
- trowel (*for digging small holes, cultivating*)
- garden fork (*for cultivating, lifting, dividing*)
- lawn rake (*for cleaning up leaves and debris*)
- Garden shears (*for pruning, trimming, deadheading*)

and I use in our gardens. They had clay pots and wheelbarrows, spoons for planting, and knives for pruning.[6]

The right tools make soil cultivation and other gardening tasks a lot easier. You'll want good, solid tools, not cheap ones that bend and break easily. You'll also want to feel comfortable using them.

Garden tools can be expensive, and there's a wide variety out there. If you're a beginning gardener, don't run out and invest in lots of exotic equipment. Start with a few good tools: a spade, shovel, rake, trowel, and garden shears. Later you may want a garden fork, hoe, hedge shears, an edger, grass shears, loppers, and long pruning shears for tall shrubs and trees.

For years I got along with just a trowel, an army shovel, garden shears, an old pair of scissors, and a rake. When I started double-digging my garden and dividing bulbs, I got a spade and a garden fork. I used to weed my gardens with a trowel, but the task is much easier with my newest tool: a Japanese *hori hori* knife. About two inches wide with one serrated edge and a slight indentation in the blade, this knife is great for getting under weeds and pulling them up. It also helps with trimming, edging, and cultivating. As you tackle more tasks in your garden, you'll find your own favorite tools.

*Choosing Garden Tools*

*When shopping for garden tools, pick up different models of the tool you want. Look for tempered steel, tools that feel solid and well-made, handles that are part of the tools or solidly riveted on, without flaws or rough spots. Hold the tool as you would in your garden. Choose one that feels good to you, suited to your size and strength. If you can't find one you like, look somewhere else.*

### Adding Compost

ADDING COMPOST TO the soil is nothing new. Thomas Tusser told people in the Renaissance:

> All gravell and sand,
> Is not the best land.

A rottenly mould [compost]
Is land worth gould [gold].[7]

You can buy bags of organic compost at the garden supply store or make your own. Since I want lots of compost for my garden, I do both. In the Middle Ages every garden was enriched by compost. Weeds, scraps, and dung from doves and farm animals improved the soil. Every kitchen garden had its compost pile, and the fork for turning and spreading compost was an essential garden tool.[8] I've set up a modern compost container in my yard near the trash and recycling bins. Going a step further, my friend Tracey has a box of earthworms that turn her kitchen scraps into rich, dark soil.

## Making Compost

YOU CAN MAKE your own compost, recycling leaves, grass, and kitchen waste into rich new soil.

- *Build a compost pile* about four feet wide and four feet high, beginning with a layer of twigs or hedge trimmings laid on the bare ground. Then alternate four- to six-inch layers of green vegetation (grass clippings and vegetable trimmings), manure, and earth, then another layer of vegetation. Keep the pile moist, sprinkling it periodically so it remains as damp as a wrung-out dishrag. Turn it with a spade or garden fork about once a week to keep it aerated.
- *Or get a compost bin* at the hardware store—an easier way to compost. Made of recycled plastic, a bin will keep the pile covered and aerated, preventing odors and speeding the composting process. In warm weather it can produce compost in eight weeks.

*Take Your Tools with You*

*Many gardeners wear their shears in a tool holster. If you want to carry more tools, get yourself a gardening belt. I have a belt that keeps my shears, knife, trowel, and gloves—all the tools I need—within easy reach. I hang the belt on a hook outside the door and put it on when I go out. Even if I think I'm just going out to water, I always seem to need my tools.*

- *All kinds of things can be composted.* From the kitchen you can compost banana peels, apple cores, carrot tops, vegetable trimmings, tea bags, coffee grounds and filters, eggshells, even stale bread (but not meat scraps or bones). Cutting scraps into small pieces helps the process. From the yard you can compost grass clippings, leaves, and deadheaded flowers (but not weeds or diseased plants).
- *To speed the process, add an activator*: manure (chicken, steer, or horse—not dog or cat), or a packaged compost starter. I've used both. Now we get all the manure we need from Bob's horse.[9]

## Tending Your Garden and Pulling Weeds

IN EARLY SPRING you'll have many small tasks as your garden comes back to life. It's the last chance to prune deciduous fruit trees, vines, and roses while they're still dormant. You'll want to rake up any leaves and plant debris, eliminating a source of disease and a hiding place for snails and slugs. I like to walk around my garden, filling a bushel basket with leaves and fallen camellia blossoms, cleaning up while discovering new buds and garden growth.

Once the winter weather recedes and warmer days return, it's time to check your drip watering system. Depending on the precipitation in your region, you may need to begin watering. Since plants need both food and water, remember to feed your roses, trees, and shrubs as well as any early vegetables.

Another vital task now is weeding. In 1580 Thomas Tusser reminded people that "digging, removing, and weeding, ye see,/ Makes herbe the more holesome and greater to bee."[10] In March, when the ground is soft from spring rains, weeds are more easily uprooted. They are also still small, more readily controlled—and wise gardeners *must* control them. Weeds choke out the other plants and absorb valuable nutrients from the soil.

This month I've been chasing dandelions in my front yard. They pop up constantly, taunting me with their bright yellow flowers. I want to catch them now, before they go to seed and multiply all over the yard. In medieval England dandelions flourished in kitchen gardens. Considered an herbal remedy and a pot herb, they were cooked in stews and prescribed for many ailments. Early settlers even brought dandelions with them to America. Today some herbalists still use dandelion root as a tonic and digestive aid, recommending it to nourish the liver and treat skin inflammations,[11] but in my garden dandelions are weeds.

What is a weed? A plant in the wrong place. The bright yellow Cape oxalis (*Oxalis pes-caprae*) plants blooming on the hillsides are beautiful. Known as Bermuda buttercups, they were brought to this country from South Africa as ornamentals. But they can easily become invasive and now grow wild all over Northern California. In my garden they are weeds, their advancing legions threatening to obliterate the tulips and daffodils.

Blackberries also grow wild in Northern California. When I moved here last fall, I was excited to see robust blackberry vines growing near the back fence. But then they began springing up all over the north side of my yard, threatening to overrun the camellias and spread their thorny branches across the brick pathway. In my garden blackberry vines are weeds I must cut back, hack out, dig up, keep under control—and when I'm not looking they pop back up again. Irascible and stubborn, they insist on taking over, so I have to root them out continually.

This spring resolve to keep weeds under control. Learn to identify the weeds in your area. A nearby nursery can give you a list and help you recognize them. Once you pull up the weeds and plant your garden, you can discourage future weeds by covering the soil with a layer of mulch. Nature abhors a vacuum, always covering the ground with *something*: grass, bedding plants, vegetables, ground cover, mulch, or weeds. In your garden it's up to you to decide. I'd rather grow flowers and vegetables, so I pull weeds and spread organic mulch throughout the growing season. The mulch keeps down the weeds while improving soil texture and conserving moisture.

## Keeping Garden Records and Journals

GARDENERS ARE ALWAYS planning—looking forward to this year's garden while looking back at the last, noting what grew well, when the last frost fell, when the first daffodils appeared. Most gardeners keep careful records of their successes and disappointments. Thomas Jefferson kept a garden book for most of his life, recording weather changes, when he sowed his first vegetables, including his favorite garden peas, and when he enjoyed his first harvests. I like to keep my "garden journal" in a box of index cards, with dividers for the months. Whenever I cultivate the soil, plant seeds, add fertilizer, or notice something new in the garden, I write it down on a card and date it. Later it's easy to look back, reviewing each month in the life of my garden. If you don't have a system yet for recording and planning your garden growth, get yourself a notebook or a set of cards and begin to chart the life of your garden.

### Planting New Seeds

IN THE RENAISSANCE, Thomas Tusser advised people in March: "Heere learne to know/ What seedes to sowe."[12] For centuries March has been the time for spring planting. What and when to plant depends on your region. In the Northeast March 17 is the traditional planting date for early peas. Once the ground has thawed, you can sow peas and other hardy vegetables—radishes, beets, cabbage, kale, and spinach. In my region we can also plant carrots, onions, turnips, white potatoes, lettuce, broccoli, and cauliflower. These are the cool-weather vegetables. Warm-weather vegetables, such as squash and tomatoes, come later. But now we can also plant sweet peas, nasturtiums, pansies, snapdragons, and calendulas.

If you'd like to grow vegetables this year, choose a good site for your kitchen garden. Ideally, the spot should be fertile, sunny (eight to ten hours for tomatoes and bell peppers), and close to your kitchen to harvest herbs and vegetables for summer meals. If you don't have enough space near your house, try growing herbs and small vegetables in pots outside your kitchen door, perhaps planting other vegetables out back. If you live in an apartment, you can grow vegetables in containers on your patio or balcony.

Take your time as you plan your new vegetable garden. Cultivate the soil or fill your containers with potting soil, then plant a few herbs and vegetables, water them, and watch them grow. You don't need to rush. It's still early in the season.

If it's too cold to sow your seeds outside, try planting some indoors in a pot near a sunny window. Cover the pot with plastic wrap until the seeds germinate. Then remove the wrap and remember to keep the seedlings moist.

Planting seeds, for me, is always an act of faith. Last year when I sowed carrot seeds in my garden, I wondered if they would ever grow. So many things could go wrong. The birds could eat the seeds before they got started, or the seedlings could perish in a late spring frost. I buried the seeds in the ground and watched. Then one day I witnessed a revelation: tiny sprouts rising in a row of bright green, a celebration of new life.

This March, since the weather has been colder than usual, all I've planted outside so far are some salad greens and onion sets. I sowed leaf lettuce and spinach in planters on the warm south side of the garden. Then I bought two bags of miniature seed onions from the garden supply store, turned over the soil, added compost, and planted them. The spring rains have done the rest. Some of nature's cycles don't take long at all. In just a few weeks the salad greens have come up, and there's already a profusion of green onions nearby, easy to harvest for salads or stir-fry dishes. I'll enjoy some green onions soon, picking others during the weeks ahead.

Onions were a favorite in medieval and Renaissance kitchen gardens, enjoyed in soups, stews, casseroles, and salads (the English upper classes began eating salads during the late Middle Ages). Among the oldest recorded vegetables, onions were cultivated by ancient Egyptians, eaten by the pharaohs, and fed to the builders of the pyramids. Along with garlic and leeks, other members of the Allium family, onions are said to aid digestion, lower cholesterol, and help prevent cancer and heart disease.[13]

You can grow your own green onions from onion sets now, either in your garden if it's warm enough or in a pot inside your kitchen window, to harvest and enjoy when you like. All you need is good soil and a bit of patience. Use green onions like chives, trimming one or two shoots with scissors to add to salads or garnish a bowl of hot soup. Enjoy them as your first fresh vegetable of the season.

## GARDEN REFLECTION

### *Beginnings and Endings*

IN OUR GARDENS March is the time to watch for new growth on shrubs and prune back the deadwood killed by last winter's frost. You can do this by looking for buds, then cutting back just above where the buds appear.

Don't give up on your plants too soon. This month I've been delighted to see hydrangeas and ferns I thought were dead begin budding forth with new life. But last winter's freeze killed the red hibiscus bush by my back window. I've watched it for weeks now and finally had to admit it was gone. I'll have to cut it back and put a hardier plant in its place.

The wisdom of Ecclesiastes says, "To every thing there is a season and a time to every purpose under heaven." Every living thing has its cycles. Some plants live for one short season. Others go through many cycles of renewal. The lives of bulbs—daffodils, hyacinths, and tulips—are perennial, with a brief season of flowering followed by long months of dormancy. After they blossom in early spring, these plants turn all their attention to their roots to produce strong bulbs for another season of flowering. If the spent blossoms are left on too long, the plants turn some of their energy into making seeds. We must deadhead them, cutting off the spent blossoms so the plants don't get exhausted trying to move their energies in two directions at once.

Deadheading spent blossoms is an ongoing garden task. This season you'll want to deadhead spent bulbs and perennials to help them grow stronger roots for next year. Later you'll want to deadhead roses and annuals to keep them blooming throughout the summer.

As I was deadheading the daffodils, my neighbor Rhonda showed me how to tie them back. After cutting back the flower and stem, take the long, slender leaves and bend them over toward the ground, keeping one leaf free to tie around the rest, and secure them in a small green bundle.

Each day, as I see new daffodils blooming in my garden, I deadhead the spent blossoms and tie back some of their leaves into these small bundles—

thank-you knots—encouraging the plant to send its energy back to the root so the bulb will grow strong and bloom again next year.

The quiet lessons in a garden teach us much about life's philosophy. The Chinese classic the *Tao Te Ching*, written twenty-six centuries ago, recorded the natural cycles of action and repose, the wisdom of yin and yang. In an ongoing cycle of beginnings and endings, the daffodils in my garden turn within in a rhythm of yin after blossoming in a burst of yang. For me tying up each plant is a ritual, a way of pausing to give thanks. It's a good lesson to return to our roots after a period of activity, to offer gratitude for the blossoming and go within for a time, honoring the alternating rhythms of nature that empower our lives as we cultivate the gardens within and around us.

# FOUR

## April: Blossoms,
## Growth, and Renewal

Whan that Aprill with his shoures soote
The droghte of March hath perced to the roote.

GEOFFREY CHAUCER
*The Canterbury Tales,*
General Prologue, ll. 1–2 (1387)

B Y APRIL OUR gardens are alive with the wonder of spring, transformed from winter grays and browns into a profusion of bright green shoots and colorful blossoms. At this magical time of year I walk around my garden with a sense of devotion, as the herbs and flowers seem joined together in a vernal hymn of praise. At such times our gardens return us to their primal meaning of paradise.

During the first April in my garden, I discovered new flowers on a daily basis. Betty Johnson had planted so many spring bulbs and perennials. To me they were all beautiful strangers. What were their names?

In Milton's *Paradise Lost*, Adam named the animals and Eve named the flowers. In that act of naming grew a relationship of understanding. "I nam'd them," said Adam, "and understood thir Nature."[1] The act of naming anchors an experience in our minds, helps us to know it more intimately. Like Shakespeare's description of the poet's art, putting an experience into words "gives to airy nothing a local habitation and a name."[2]

When I discovered new flowers, I asked friends and neighbors what they were, looked them up in the *Western Garden Book*, researched their individual lives and stories. I wanted to learn their botanical names. I'd rather call the beautiful ground cover in the inner courtyard *Ajuga reptans*, not "bugleweed." Its common name I found unsuited to this charming plant with copper-tinged leaves and spires of purple flowers, for it was not a weed, and the Latin name was much more descriptive. *Reptans* refers to the way it spreads, its runners creeping along the ground like reptiles.

Botanical names help identify a plant, preventing confusion. Often the same plant has many common names—the yellow *Oxalis pes-caprae* in my yard is called both Bermuda buttercup and wood sorrel. Or sometimes many plants have the same common name. *Arctotis, Dimorphotheca,* and *Osteospermum* are all called African daisy. To get the one you want at a nursery, you'll need its botanical name.

Our system of botanical naming was established in the early eighteenth century by the Swedish explorer and botanist Carolus Linnaeus. The first part of the name is the plant's genus, the second its species. The third name, if listed, refers to the expert who classified the plant. Many plants' names include "L." for Linnaeus, who named thousands of species. One of these is rosemary, or *Rosmarinus officinalis* L.

A plant's species name often provides descriptive information. *Officinalis* or *officianale* means "from the apothecary's shop," indicating that the plant was used medicinally. Many familiar herbs have this name. Other species names refer to color (white = *alba*, red = *rubrus*, purple =

*Laurel and Myrtle, and
what higher grew
Of firm and fragrant leaf;
on either side
Acanthus, and each
odorous bushy shrub
Fenc'd up the Verdant
Wall; each beauteous
flow'r,
Iris all hues, Roses, and
Jessamin
Rear'd high thir flourisht
heads between, and
wrought
Mosaic; underfoot the
Violet,
Crocus, and Hyacinth
with rich inlay
Broider'd the ground,
more color'd than with
stone
Of costliest Emblem.*

JOHN MILTON,
*Paradise Lost*, IV,
ll. 694–763 (1674)

*purpureum*), foliage (large leaves = *grandifolia*, many leaves = *mille-folum*), or the region where it grows (near the ocean = *maritima*, in the mountains = *montana*). When we know these things, names like that of the medieval white rose, *Rosa alba*, don't seem so mysterious. If you haven't learned the botanical names of your plants, I encourage you to begin. From now on I'll provide botanical names in my garden descriptions.[3]

In April, as gardens blossom with new life, many celebrations have hailed the return of spring. The jests of April Fool's Day on April 1 echo earlier pagan new year's festivities. The medieval and Renaissance English celebrated the feast of St. George on April 23, wearing the first red roses of the year. Late March or April ushers in the Easter season, marking an end to the dark days of winter and Lenten deprivation. In 525 C.E. the Catholic Church decreed that Easter would be celebrated on the Sunday following the first full moon after the vernal equinox (March 21), so this holiday is a movable feast, occurring on different dates each year.

Christian Easter celebrations center on the resurrection of Christ and the promise of eternal life. But many Easter traditions, including the name itself, have earlier connotations. Eostre or Ostara was the Anglo-Saxon goddess of spring, and her symbol was the rabbit. Decorated eggs were given throughout Western Europe, as well as in ancient Persia, Russia, and Bulgaria, to celebrate the return of life at the vernal equinox. Even without looking at this rich history, we would know that rabbits and eggs are fertility symbols, and, however we celebrate this season of renewal, our gardens in April reaffirm for us the miracle of life.[4]

## GARDEN GROWTH

EACH MONTH IN my garden, nature's shifting cycles become more apparent. In April, as the last daffodils and grape hyacinths (*Narcissus* and *Muscari*) disappear, tulips and irises take their place, while California poppies (*Eschscholzia californica*) spring up all over the garden. Now California's state flower, the golden poppy was once used by Native Americans as a sedative and painkiller for toothaches. When the Spanish conquistadors arrived and saw the poppies blooming in the distance, they thought the hills of California were covered with gold.[5]

In early April tulips are my garden's star performers,, raising their cup-shaped blossoms of crimson, rose, pale peach, cream, yellow, purple, and lilac. Before they open they look like painted Easter eggs. Tulips (*Tulipa*) came from Persia and Syria, and were imported from Turkey in the sixteenth century by Dutch traders. Throughout the Renaissance the Dutch raised and cultivated tulips, until these bulbs became a sensation all over Europe, resulting in highly inflated prices and bidding wars during the seventeenth century.[6]

Tiny white starlike blossoms of onion lily (*Allium triquetrum*) bloom throughout the garden. Although some consider them weeds, I find these wild onions with their Easter grass foliage a perfect complement for the tulips. While weeding I discovered that they definitely smell like onions. Their leaves and bulbs are even edible.

By late April, as the tulips fade, the irises raise their bright standards around the yard. Dutch irises bloom in violet tipped with gold, joined in a few days by others in purple, white, and butter yellow. In the next few weeks they are followed by bearded irises in rainbow colors: purple, blue-violet, lilac, apricot, yellow, peach, rose, maroon, white, pale green, and one such a deep purple it seems almost black.

Named for the Greek goddess of the rainbow, the iris grew throughout medieval and Renaissance Europe. There are many varieties, including the yellow *Iris pseudacorus* L., adopted by the French King Louis VII as his emblem, which became the French fleur-de-lis. The white iris became the symbol of the Virgin Mary, and the blue iris in medieval England

*Faire Daffadills, we weep to see*
*You haste away so soone.*
ROBERT HERRICK,
To Daffodills,
ll. 1–2, from
*Hesperides* (1648)

*You are a Tulip seen today,*
*But (Dearest) of so short a stay;*
*That where you grew, scarce man can say.*
ROBERT HERRICK,
"A Meditation for his Mistresse,"
ll. 1–3 (1648)

was prized for its flowers and its varied uses. Its foliage was strewn on floors and woven to cover chairs and to thatch roofs; its roots were made into ink, scents, and air fresheners; its flowers were blended with alum to make green dye; and its juice was used as a household remedy to lighten skin and clean the teeth.[7]

Along with the irises, red and blue windflowers (*Anemone blanda* and *Anemone coronaria*) bloom around the front gardens, joined by *Ranunculus asiaticus* in crimson, rose, white, tangerine, and yellow, and the last fragrant *Freesia* in violet, red, and gold.

Behind the bulbs, in a shady area along the courtyard fence, clusters of bright cineraria (*Senecio hybridus*) glow like amethyst gems. White sweet alyssum (*Lobularia maritima*), blue lobelia (*Lobelia erinus*), and nasturtiums (*Tropaeolum majus*) in bright citrus colors have reseeded themselves, adding ground color throughout the garden. Nasturtiums are a favorite flower of mine. They're not only good to look at but good to eat. Their spicy sweet flowers are wonderful in salads or nouvelle cuisine and their seeds are often used in place of capers.[8]

*Such comfort as do lusty young men feel When well-apparelled April on the heel Of limping winter treads.*
WILLIAM SHAKESPEARE, *Romeo and Juliet,* I.2, ll. 24–26 (1595)

I'm greeted everywhere by new spring blossoms. The first buds have appeared on the roses, the perennial candytuft (*Iberis sempervirens*) beside the front gate is a mass of snowy white flowers, and the French lavender growing along the front wall is covered with fragrant purple blooms. Lavender comes in many varieties, including French lavender (*Lavandula dentata*) and English lavender (*Lavandula officinalis*), introduced to Britain by the Romans and used from the Middle Ages to the Renaissance in perfumes, bath oils, medicines, and foods. Lavender conserve, a confection of lavender flowers and sugar, was a favorite of Queen Elizabeth I.

In the herb garden pale blue blossoms adorn the rosemary I planted last fall. From the Middle Ages through the Renaissance, *Rosmarinus officinalis* L. was used for cooking, strewing, and a variety of medicines. It was also believed to improve memory. Students in ancient Greece wore rosemary sprigs in their hair when they studied.[9] When I told my seminar students this, they laughed, but aromatherapists still use herbal scents to stimulate our brains and renew our spirits.

Azaleas and camellias bloom in the front courtyard and along the north garden walk in shades of pink and rose. In the backyard the wisteria arbor is graced with grapelike clusters of fragrant blossoms (*Wisteria sinensis*), bright green leaves bedeck once-dormant trees and shrubs, rhododendrons bloom around the perimeter, and the tea tree (*Leptospermum scoparium*) outside my study window is filled with magenta blossoms that look like tiny roses. This plant got its common name in the eighteenth century, when the British explorer Captain Cook served his crew a tea brewed from the leaves to prevent scurvy during long voyages in the South Pacific. The tea was effective, but its taste was bitter and medicinal.[10]

In the friendship garden between my house and Rhonda's, colorful clusters of red harlequin flowers (*Sparaxis tricolor*) and deep rose Watsonia blossoms (*Watsonia pyramidata*) from South Africa bloom among the irises and California poppies. Along the south wall the onion sets and salad greens I planted last month are growing well, and the warmer April days invite me to plant more herbs and vegetables.

## GARDENING AS SPIRITUAL PRACTICE

### *Discovering New Growth*

FROM THE GARDENS around us to the gardens within us, April brings new energies and new beginnings. Like the seeds we planted in our gardens last month, our seeds of personal renewal should be showing signs of life. Now is the time to evaluate this new growth in our inner gardens.

If the new goal or direction you chose last month has added valuable dimensions to your life, you'll want to continue cultivating it with positive action. If your seeds of renewal failed to germinate, you'll want to evaluate your goals and the ground you set them in, making some necessary changes. If your seeds are growing so fast they've filled up your life with excessive activity, you'll need to do some thinning out.

## Questions to Consider

THIS WEEK GIVE yourself a few minutes to answer these questions, determining what you'll need to move forward in this joyous season of growth.

- Did your seeds from last month germinate and grow? Are you happy with the results?
- If so, what steps can you take to continue moving toward your goal?
- If your seeds didn't germinate, what went wrong? Was it the goal? The timing? The atmosphere? What can you do about this?
- If your seeds are filling up your life with too much activity, where can you do some thinning?
- What have you learned from this experience? What would you do differently next time?
- What new seeds and actions can you plant this month? Write them down and put the first action on your calendar.

## Sowing New Seeds: Making Mindful Choices

APRIL IS A wonderful time to begin anything. The warmer days, bright green leaves, and glorious blossoms bring us new energy and enthusiasm. But sometimes we let the energies of spring move us in too many directions at once. In this season of new growth, we need to make mindful choices. Wise inner gardeners resist impulsiveness and consider the patterns of their lives before sowing seeds for more new beginnings.

We sow seeds in our personal gardens whenever we begin new projects or new activities. But the greatest stress for many people is too much activity. This is especially true where I live, in California's Silicon Valley. Some people blame computers, downsizing, multitasking, or new economic conditions. But whatever the reason, more and more of us are caught up in a rush of escalating demands and incessant activity.

Regardless of occupation, when we rush from one activity to the next we leave little time for reflection and renewal, and this frantic cycle only reinforces itself. As an accelerated pace becomes the norm, we overschedule ourselves, even in our private lives. Our lives' gardens become overcrowded. Too rushed to reflect, we often make foolish choices, filling our lives with too many activities, surrendering to stress, denying our deepest values.

> *Oh, what a thing is man!*
> *How farre from power,*
> *From setled peace and*
> *rest!*
> *He is some twentie sev'rall*
> *men at least*
> *Each sev'rall houre."*
> GEORGE HERBERT,
> "Giddinesse," ll. 1–4,
> from *The Temple*
> (1633)

## Questions to Consider

DON'T LET YOUR inner garden become overcrowded by too much new growth at once. Before you commit yourself to more than you can handle, take time to ask

- Is this new challenge right for me now? Does it make sense in the larger pattern of my life?
- Does it express my deepest values and lead in the direction of my dreams?
- Am I prepared for it? Do I have the necessary skills? The time?
- Would this new activity make my life more harmonious—or just busier?
- When I visualize myself doing this, how do I feel?

Listen carefully. Your heart will often tell you more than your mind can know.

This month I was offered an administrative position at work. The job would probably mean more money. It would also mean long hours, lots of paperwork, meetings, less time to teach and write. It would be a wonderful opportunity for someone who was passionate about administration but only a detour for someone who was not.

The King James Bible tells us that "there are diversities of gifts, but the same Spirit."[11] Each of us has different gifts, different talents. Some are artists; others are healers, teachers, researchers, merchants, inventors, engineers, civil servants, parents, and community leaders. In Renaissance England people believed that one of life's most valuable lessons was recognizing our gifts, cultivating them to the best of our ability, and using them to love and serve the world. I still believe this.

## Questions to Consider

- ❧ What are *your* special gifts and talents?
- ❧ Do *your* current choices and commitments enable you to use your gifts to love and serve the world? If so, you have a perennial source of joy in your life. If not, why not? What actions can you take this month to cultivate this part of your inner garden?

## Thinning Your Seedlings

So let thy Vines in
Intervals be set,
But not their Rural
Discipline forget:
Indulge their Width, and
add a roomy Space,
That their extreamest
Lines may scarce
embrace.

JOHN DRYDEN,
Virgil's Georgics, II,
ll. 386–89 (1697)

EVEN WHEN WE'RE mindful about beginning new projects, we often get more than we bargained for. As our seeds begin to germinate, we often have to thin them. Today I thinned some of the mesclun greens in the south garden. The seedlings were crowded together in clumps. It seemed a shame to pull up some of the tiny things, but if they're too crowded, they won't grow well.

We *can* have too much of a good thing: too many projects, activities, commitments. Our gardens remind us that less is more: we thin out some growth for the rest to flourish. After bringing in the lettuce thinnings to add to tonight's salad, I went into my study and began sifting through piles of paperwork, discarding some projects to concentrate on others. It's the same principle. We all do better when we have room to grow.

## Questions to Consider

- ❧ How balanced is the growth in your life's garden?
- ❧ Do you have room to grow?
- ❧ If not, what activities can you thin out to leave more space for reflection and renewal? Write these down in your journal or on an index card.
- ❧ Think of ways to delegate the tasks you've thinned out. Who else can do them?
- ❧ Take a deep breath and see yourself delegating, creating wider margins in your life.
- ❧ Play a game: Find a way to make your victories visible whenever you delegate or release an unwanted task.

This month I got myself some green adhesive dots from the stationery store to put on my calendar whenever I delegate something: a symbol of the "green space" I'm clearing for healthier growth. Since I've measured my progress for years by "points" at school or at work, I decided to use this conditioning to cultivate my inner garden.

The funny thing is that now I *look* for ways to delegate, just to win green points, a private reward no one else even sees. I used to do lots of my own photocopying, even while I was department chair or faculty senate president. It *seemed* more efficient, but that was an illusion. Since I've been delegating my photocopying to student assistants, I don't feel so overwhelmed at work. All that photocopying time adds up. I now have more margins, more room to grow.

## GARDEN TASKS

APRIL WEATHER IS defined by mutability. One day my garden is tossed by rough winds; the next day it rains; the next brings blue skies and sunshine.

The warm, sunny days tempt us to begin setting out flowers and vegetables, but April frosts and snowstorms can still surprise some of us. Mindful of nature's cycles, wise gardeners take nothing for granted. You'll need to watch the weather conditions in your area to know when to plant, when to protect tender plants from frost, and when to begin watering.

*Sweete April showers*
*Doo spring Maie flowers.*
THOMAS TUSSER,
*Five Hundred Points*
*of Good Husbandry*
(1580)

## Watering, Weeding, and Feeding

AS THE WINTER rains taper off in Northern California, it's time for me to begin watering in earnest, paying special attention to new plants. I've put soaker hoses around the north yard and spent lots of time watering the front gardens. I'd like to install a drip system there, not only to conserve water but to benefit the roses, which dislike getting their leaves wet. Many plants—roses, dahlias, squash, beans, and tomatoes—get powdery mildew and other diseases when their foliage remains damp.

*Yoong plants soone die*
*That growe too drie.*
THOMAS TUSSER,
*Five Hundred Points*
*of Good Husbandry*
(1580)

I've also been doing a lot of weeding: pulling up invasive grasses, dandelions, and oxalis, and discovering new flowers and emerging growth in the process. Just yesterday along the front border I found a white pansy with a purple-and-gold center (*Viola wittrockiana*). To discourage weeds, help amend the soil, and keep it moist, I've been spreading compost around as mulch. Other necessary tasks are deadheading spent blossoms from early bulbs, picking up fallen camellia blossoms to prevent disease, and thinning early vegetable seedlings.

As our plants begin their season of active growth, it's time to feed them. This week I fed the lemon and dwarf peach trees and hammered rose food stakes among the roses. Releasing fertilizer to their roots for six weeks at a time, these stakes keep the roses blooming happily. Betty Johnson left a package of rose stakes for us last fall when we moved in. I also have a box of special acid food to give to the camellias, azaleas, and rhododendrons when they finish blooming. But for now I'm still enjoying their colorful blossoms. (Botanically, camellias are *Camellia japonica.* Azaleas and rhododendron are species of *Rhododendron.*)

## Pest Control

AS WARMER WEATHER returns to our gardens, so do many garden pests. Snails, slugs, and aphids feast on new garden growth and can decimate our plants almost overnight. Early this month I discovered aphids on the rosebuds and sprayed them with insecti-

*For want of seede*
*Land yeeldeth weede.*
THOMAS TUSSER,
*Five Hundred Points of*
*Good Husbandry*
(1580)

cidal soap. The tiny cowboys of the insect world, ants herd aphids like cattle, defending them from predators and milking them for their sweet excretions or honeydew. An army of ants can march in with a battalion of aphids and garrison them on your plants in record time. One way to control aphids is to keep ants away with barriers of sticky pest control paste, available at local nurseries. You can also attract natural predators: ladybugs, soldier beetles, lacewings, and tiny parasitoid wasps that attack aphids and lay eggs inside their bodies. These beneficial insects usually appear a few days after initial aphid infestations. Since I don't want to use toxic products in my garden, I spray an insecticidal soap to kill the aphids on contact without harming the plant, the soil, or any beneficial insects that come by later.[12]

The most notorious pests in my garden are slugs and snails. This spring they have been on a rampage, eating daffodils, irises, emerging dahlias, and some of the tender herbs in my kitchen garden. I surrounded the herbs with diatomaceous earth, tiny sharp fragments of marine animals, which discourage snails and slugs by piercing their bodies, but that doesn't seem to be working. Since I don't want to use toxic snail bait—it poisons the earth and could also kill our dog and cat—I sought other options.

When my neighbor Rhonda finds snails in her yard, she just picks them up and hurls them into the street. I'm basically a nonviolent person, but I began following her example, gingerly picking up the snails first with my trowel, then with my garden gloves. Finally, one morning when I found a fat snail feasting on an iris, I boldly picked it up with my bare hands, hurling it into oblivion.

But still there are more snails than I can handle. This year's rains have produced a record infestation. At the hardware store I overheard a silver-haired lady echo lines from Jacobean drama, demanding "the most powerful poison available" to rid her garden of snails. Unwilling to resort to such extremes, I put copper foil around my containers of salad greens. The foil is supposed to repel snails by giving them a slight electrical shock. At least it makes the pots look festive. After hearing how foolish gastropods will drown in tins of beer set out at night, I bought three green plastic traps, set them in the ground, and poured in the beer. The next morning the traps were filled with moribund party animals. For the past few weeks I've gone through gallons of beer and spent many mornings cleaning up after the fatal parties, dumping all the dead slugs and snails into the compost pile, which has begun to smell like a brewery.

## Planning Your Spring Garden

THE WARM APRIL days make us believe anything is possible. Gardeners rush out to local nurseries, eager to plant their spring crops. The annuals and spring vegetables look so appealing, it's easy to buy more than we'd planned. But it makes sense to slow down, to assess our gardens' possibilities before running through the nursery in a fit of spring fever.

Getting to know your garden means learning your plant hardiness zone. The U.S. Department of Agriculture publishes a Plant Hardiness Zone Map (see page 277), which divides our country into eleven zones, based on average winter temperatures. These range from the coldest areas, zones 1 and 2 in Alaska (from below −50 to −40 degrees Fahrenheit) and zone 3 in the Upper Midwest (−40 to −30 degrees), to the warmest tropical zone 11 in Hawaii (above 40 degrees). My region in Northern California is in zone 9, with winter lows between 20 and 30 degrees. Knowing your hardiness zone is important when planting shrubs, trees, and perennials that live through the winter. Also important, especially when planting annuals, is your area's last frost date, so you can safely set out tender flowers and vegetables. Check with your nursery or agricultural extension agent about local frost patterns.

To choose plants that will thrive in your garden this summer, it's important to know your area's heat extremes. After two decades of research the American Horticultural Society has produced a heat zone map, dividing the United States into twelve zones based on the average number of days a year temperatures rise above 86 degrees Fahrenheit (30 degrees Celsius). My area is heat zone 4, with fourteen to thirty days a year above 86 degrees. You can find out your heat zone by contacting the American Horticultural Society.[13] Some nurseries have begun labeling plants with heat-tolerance designations.

You'll also want to know your garden's microclimates: the warmest and coolest sections, which areas get the most sun, and which have partial shade. Like people, plants have different preferences. Salad greens, onions, and thyme prefer partial shade, while tomatoes, squash, and peppers require full sun and lots of warmth. The southern part of a yard is usually the warmest. Walls or cement radiate even more warmth. I plant my tomatoes in containers along our sunny south garden walk, which has heat reflected off the house's stucco wall and the cement-and-stone walkway. Shady areas will remain cooler, and windy areas will dry out more quickly. If your house is on a hill, areas near the bottom can trap cold air while the top remains warmer. Thomas Jefferson's fruit trees at Monticello often escaped the frost damage experienced by his neighbors. Monticello, of course, means "little hill," a setting that gave him a delightful view and helped protect his orchard from frost.[14]

## Planting Your Vegetable Garden

THIS MONTH MANY of us plant our vegetable gardens. The exact timing will depend on local conditions. As soon as the ground can be worked, we can plant cool-weather vegetables: peas, beets, radishes, potatoes, onions,

*In March and in April,*
*from morning to night*
*In sowing and setting,*
*good huswives delight.*
THOMAS TUSSER,
*Five Hundred Points*
*of Good Husbandry*
(1580)

and salad greens. Last month I planted onion sets, mesclun greens, and spinach. Later, when the ground warms up and the danger of frost has passed, we can plant warm-weather vegetables: beans, carrots, corn, cantaloupes, cucumbers, eggplant, squash, peppers, and tomatoes, as well as annual herbs. You'll need to wait to plant these until daytime temperatures average 70 degrees Fahrenheit and soil temperatures reach 60 degrees. I got a soil thermometer at the local hardware store and have been taking my soil's temperature, which is already over 60 in the south garden.[15] There I've set out three tomato plants. I also sowed more mesclun greens and bok choy in containers.

## Growing Tomatoes

TOMATOES NEED FULL sun and lots of warmth, but they're relatively easy to grow, and nothing tastes like vine-ripened tomatoes. I set out tomato plants when the days first warm up, so I can begin harvesting them as soon as possible. Your local nursery will have varieties that grow best in your area. The labels will tell you the following:

- whether they grow well in containers
- what diseases they are resistant to (V = verticillium, F and FF are different strains of fusarium wilt, N = nematodes, and T = tobacco mosaic virus)
- whether they are determinate (growing as a bush three to four feet tall and producing all their fruit at once) or indeterminate (continuing to grow and bear fruit throughout the season)
- how many days they'll take to mature (tomatoes usually range from fifty-five to ninety days)

## Checklist for Setting Out New Plants

HERE ARE A few pointers for setting out new plants in your garden:

- Turn the pot or six-pack upside down, tapping the pot or squeezing one cell of the six-pack with one hand while placing the plant stem between the index and middle fingers of your other hand.
- The plant should slip out into your hand. If it's difficult to remove, run a knife blade around the edges of the pot or cell.
- Once you've removed the plant, examine the roots. If they are in a tangled mass at the bottom, tear off the matted part.
- Loosen the roots at the bottom of the root ball.
- Place the plant in a spacious hole, and fill the area around it with additional soil.
- Press down on the soil to remove air pockets.
- Water the plant in well, soaking the soil until a small puddle forms around the plant. But be gentle—don't blast the young plant with the hose.
- Add a mild fertilizer, and mulch as needed.
- Remember to water new plants often to help them get a good start in their new environment.

## Sowing New Seeds

I LOVE FRESH green beans, so this year I decided to plant some beside the north garden fence. As an experiment I ordered some special French green beans, real *haricots verts*, from Shepherd Seeds in Connecticut. I thought it would be great fun to serve them to my friend Catherine, who comes from Marseille and loves *haricots verts*. On April 20 I dusted the seeds with an inoculant, a culture of nitrogen-fixing bacteria that helps legumes produce more nitrogen from the soil.[16] Then I sowed them in the ground, counting on the April showers to do the rest.

I was a little nervous about planting these beans. What if the birds or snails ate them all? Gardening always involves risks. We sow our seeds and wait. Sending away for these special beans increased my anxiety because they are so rare, so I made a backup plan: if these beans don't make it, I'll get regular American beans—Blue Lake or Kentucky Wonders—both of which I've grown before. For insurance I planted some Kentucky Wonder beans five days later by the trellis near the north garden gate. That way, even if some beans don't make it, the others will. I love to see their vines twining up fences and producing all those tender green beans.

## Planting Spring Flowers

WHEN CONDITIONS ARE warm enough for beans and tomatoes, you can also sow and set out summer annuals. Your local nursery will have a colorful supply of asters, cineraria, coleus, forget-me-nots, nasturtiums, petunias, lobelia, snapdragons, salvia, sweet alyssum, zinnias, and other varieties. They will also have dahlias, gladioluses, and other summer bulbs, which you can plant at the same time. When the shoots emerge from summer bulbs, begin fertilizing them, and when you deadhead spring bulbs, fertilize them too, to help them bloom again next spring. This time of year I walk around with a bag of bonemeal, digging in spoonfuls as I go.

## Planting Your Herb Garden

I LOVE FRESH herbs and have grown them for years, on sunny rooftops, patios, or windowsills when I didn't have a yard. It's such a pleasure to snip fresh herbs when I'm preparing meals. Sprinkled in salads, sauces, and roasted foods, herbs add spice and savor to our lives.

My friend Jane Curry, an international relations professor specializing in Eastern Europe, gives wonderful dinners where professors, students, journalists, diplomats, and their families share stories and enjoy Jane's international cuisine. This year she had a Polish Easter dinner, with tender new potatoes that had everyone going back for seconds. Her secret: fresh produce and fresh garden herbs.

This month I set out basil and dill (*Ocimum basilicum* and *Anethum graveolens*) along the south wall and planted thyme (*Thymus vulgaris*), chives (*Allium schoenoprasum*), parsley (*Petroselinum crispum*), and rosemary (*Rosmarinus officinalis*) outside our front gate. Herbs are easy to grow, thriving in less than perfect soil and requiring minimal room. Native to the Mediterranean, rosemary and lavender can even survive with little water. Herbs do well in containers and can be grown among vegetables or in flower borders. I like to mix herbs with my spring bulbs and summer annuals, enjoying the delicious scents and colors of their foliage, the gray-green lavender and the golden and plum-green leaves of pineapple and purple sage. Some people grow herbs in pots just outside their kitchen doors for convenience. In regions with freezing winters, they bring their herbs in to overwinter inside.

Herbs had many medicinal uses in the Middle Ages and the Renaissance. Without hospitals and drugstores, people turned to their gardens to heal colds, sore throats, upset stomachs, and other ailments. They used garlic as a tonic and antiseptic; dill as a sedative and cure for indigestion; thyme for coughs and colds; marjoram to treat aching joints; rosemary to improve memory, and relieve stiff necks and aching joints; lavender for headaches; mint for colds, anger, and indigestion; sage as an all-purpose remedy; and parsley to remedy indigestion and migraines.[17]

### Polish New Potatoes

*These new potatoes are delicious and surprisingly easy.*

1 *pound small red new potatoes, well scrubbed*

2 *tablespoons fresh dill, finely chopped*

3 *tablespoons butter or butter substitute*

*Boil the potatoes in water until tender, from 10 to 20 minutes depending on their size. Drain in a colander and place, while still warm, in a heatproof glass bowl. Toss with fresh dill and butter. Salt and pepper to taste. Serve with any main course. Especially good for Easter dinner.*

## Garden Discoveries: Wandering Bees

ONE DAY IN mid-April I found a mass of bees clinging to the fence in the corner of our front courtyard. Watching hundreds of them buzzing and flying around, I was first alarmed, then intrigued. Medieval and Renaissance gardeners raised bees in wicker hives woven like baskets, prizing them for their honey.

The sweetness of honey was eulogized by generations of Irish and

Welsh poets, celebrated in the biblical Song of Songs. Anglo-Saxon war-riors drank mead, brewed from fermented honey. Benedictine monaster-ies kept bees for honey and wax, which they made into candles. In the Middle Ages honey was not only a sweetener for meats, cakes, pies, and jams but also a medicine to treat coughs, indigestion, and skin irrita-tions. Today honey is still used by herbalists to calm the mind, renew energy, and relieve allergies, and to heal colds, coughs, indigestion, and minor skin wounds.[18]

> Rosemary in the spring
> time
> To growe south or west.
> THOMAS TUSSER,
> Five Hundred Points
> of Good Husbandry
> (1580)

Bees are wonderful pollinators in our gardens, and many familiar herbs and flowers, including lavender (*Lavandula officinalis*), lemon balm (*Melissa offici-nalis*), rosemary (*Rosmarinus officinalis*), thyme (*Thymus vulgaris*), red clover (*Trifolium pratense*), California poppies (*Eschscholzia californica*), and sweet alyssum (*Lobularia maritima*), attract them. Bees are very sensitive to toxins. Sadly, most wild honeybees have been killed off in recent years by pesticides and pollution. But if we are careful other varieties, including mason bees, orchard bees, and bum-blebees, will still be around to pollinate our gardens.

The hundreds of bees who camped on my courtyard fence moved on in a couple of days. But other bees are happily buzzing around the lavender, rosemary, poppies, sweet alyssum, and other fragrant flowers in my garden.

## GARDEN REFLECTION

### Sacred Spaces: Lectio Divina

LAST FRIDAY AFTERNOON I had coffee on campus with two fascinat-ing people, Lillas Brown from the University of Saskatchewan and the South African leadership consultant Ketan Lahkani. "America doesn't have a culture. It has an economy," said Ketan provocatively. When asked to explain, he said that while we've created a booming economy, the

American mainstream has lost touch with spirituality, nature, family, and community. Much of what we value has been packaged, commodified, reduced to numbers.

As I drove home the news on the radio seemed to confirm this, for it was focused on numbers, rising interest rates and the ups and downs of the stock market. I spent the golden hour of vespers in my garden, where numbers do provide useful information. The thermometer on the south wall read 68 degrees, the soil thermometer 60—important when it comes to setting out tomatoes and germinating seeds. But the peace I find in the garden transcends numerical measure. How can one assess the value of trees that enclose the backyard in a varied tapestry of green? The new leaves on the rose of Sharon that only last week looked dead to the world? The glorious spring blossoms that make the front yard a canticle to spring? The sacred silence among the herbs and flowers?

Our gardens offer a welcome relief from the reductive values of popular culture. The contemplative time I spend in my garden makes me more present, more reverent, more whole. Medieval monastics considered private time for reflection and inspiration as important as the ritualized prayer of the Divine Office, the *Opus Dei* or work of God. *Lectio divina*, literally "the reading and contemplation of Scripture," has come to mean time spent in communion with the sacred.[19] *Lectio* to the medieval Saint Bonaventura meant meditating upon the world of nature, a practice continued in cloisters and meditation gardens throughout the Renaissance, embraced by the seventeenth-century Metaphysical poets, and taken to heart by gardeners from their time to our own.

For me time in the garden is *lectio divina*, sacred time, time spent witnessing the glory of creation. Time in the garden restores my capacity for education, inspiration, and creativity. It makes me more aware of nature's cycles, more open to love and embrace those essential qualities of life that transcend our ability to measure.

# FIVE

# May: Celebration, Cultivation, and Contemplation

Hail bounteous *May* that dost inspire
Mirth and youth and warm desire!

JOHN MILTON,
"Song: On May Morning,"
ll. 5–6 (c. 1629)

FOR CENTURIES GARDENERS, poets, lovers, and saints have praised the glorious month of May. By now the skies are bright blue, new crops are springing up in the fields, and the world is alive with flowers and song.

In the Middle Ages the beauty of the natural world in this "joly tyme of May" enticed even the poet Chaucer from his study. When he saw the flowers bloom and heard the birds, or "smale foules," sing, he had to be out enjoying the splendor of the season. In the Renaissance, Shakespeare's works were filled with praise for the month of May. Seventeenth-century devotional poets likened this month to the grace of God, which quickens the soul with a joyous renewal of life.

On May Day, from the Middle Ages through the Renaissance, young couples would gather flowers in the countryside at night, returning at dawn to decorate themselves, their homes, and their neighborhoods in rituals derived from the earlier fertility feast of Beltane. People celebrated "the merry month of May" by dancing around the Maypole, visiting May fairs, crowning a lovely maiden as queen of the May, and, often, falling in love.

May Day celebrations continued in England until 1644, when the Puritans outlawed them as pagan superstition.[1] In 1660 they were were revived with the Restoration. Even now some towns in England and America celebrate the first of May with flowers, fairs, and May baskets. American high school proms, with their corsages and queens, have probably descended from this early rite of spring.

> *In the joly tyme of May,*
> *Whan that I here the*
> *smale foules synge,*
> *And that the floures*
> *gynne for to sprynge*
> *Farwel my stodye, as*
> *lastynge that sesoun!*
> GEOFFREY CHAUCER,
> *The Legend of*
> *Good Women,*
> General Prologue,
> ll. 36–39 (c. 1386)

To supplant all the pagan celebrations, the medieval Catholic Church dedicated the month of May to the Virgin Mary and celebrated the feast of Saints Philip and James on May 1. Many other religious feasts occurred this month, including Rogation Sunday, five weeks after Easter, when priests and their parishioners would walk through the countryside blessing the fields and crops, followed by the Feast of the Ascension a few days later. Pentecost, the seventh Sunday after Easter, was celebrated as Whitsuntide, with three days of feasting, drinking, dancing, and games. Today in England most of the feasting has disappeared and the Monday of Whitsuntide has become merely Bank Holiday Monday.[2]

## GARDEN GROWTH

IN MY GARDEN May marks the transition between the exuberance of spring bulbs and the warm profusion of summer. Early this month some bearded irises were still blooming in shades of yellow, purple, and white, along with more exotic mixtures: ruffled lilac and white, burgundy and gold, and pale café au lait tinged with violet. In a few weeks I was deadheading the spent iris stalks, enjoying the last of their fresh spring fragrance, and watching time's subtle changes in my garden's design.

As the last irises have faded, the herb garden has settled into a carpet of low-growing herbs and flowers: pineapple sage, purple sage, thyme, rosemary, chives, and oregano, their subtle greens accented by white alyssum, blue lobelia, and red and gold nasturtiums. The yellowing foliage from this spring's daffodils, tulips, and onion lilies recalls an autumnal landscape, resembling diminutive cornstalks and piles of hay. Yet in front of them has emerged a colorful chorus of pansies, raising their velvet faces in shades of burgundy, rose, purple, cream, burnt orange, and gold.

Beside our front garden gate I discovered a patch of English daisies springing up in the grass, reminding me of a medieval flowery mead. I smiled, thinking of Chaucer's expansive praise of daisies in *The Legend of Good Women* — "these floures white and rede," which he loved most of all the spring blossoms.[3] The English daisy (*Bellis perennis*), a meadow flower that prefers moist soil and filtered sun, was popular throughout the Middle Ages. Grown in gardens and lawns, it was associated with innocence, woven into garlands and daisy chains. The name *daisy* came from the Anglo-Saxon *daegeseage* (day's eye), because these flowers resemble tiny, radiant suns.[4]

The daisies' shape is echoed around my garden by the amethyst sunbursts of cineraria blooming in the shade behind them and the large yellow gerbera daisies (*Gerbera jamesonii*) blossoming in the inner courtyard. Raising their heads in exultation, all these tiny sunbursts magnify the sunshine days of May.

As spring flowers fade, their summer successors appear. Throughout the front gardens the first flush of California poppies is over. While cutting back their dried stalks in the friendship garden, I discovered the rising spires of liatris (*Liatris spicata*) that will become tall plumes of violet blos-

> How fresh, O Lord, how
>    sweet and clean
> Are thy returns! ev'n as
>    the flowers in spring;
> To which, besides their
>    own demean,
> The late-past frosts
>    tributes of pleasure
>    bring.
> Grief melts away
> Like snow in May,
> As if there were no such
>    cold thing.
>        GEORGE HERBERT,
>        "The Flower"
>        ll. 1–7 (1633)

soms. Just yesterday the first lavender and pink blossoms appeared on the sweet peas Rhonda planted along the fence between our yards. Last week, when Ron and Rhonda went away on vacation, I watered the sweet peas and put bamboo stakes behind them to help the tiny plants wind their way up the fence. I was amazed at how soon they took hold and began climbing the stakes. Vining plants—sweet peas, garden peas, and beans— wind their way up poles, trellises, and fences with a kind of intelligence botanists call thigmotropism, a word drawn from the Greek *thigma* (to touch). Some plants can wrap their tendrils around a support more than once in under an hour.[5]

By now all the azaleas and camellias have stopped blooming in the front courtyard and along the north garden walk. The wisteria blossoms in the backyard are gone, but the peach blossoms have become small green fruit, the hydrangeas are beginning to bloom, and there are bright orange blossoms on the rhododendron I repotted when we moved in last fall.

The vegetable gardens are showing promise. The first yellow blossoms have appeared on the tomatoes, bok choy seedlings have joined the lettuce and spinach greens, and the first French beans are raising their heads above the surface.

But for me the roses are the greatest gift of May. My garden has hybrid tea roses: red Mister Lincoln and Chrysler Imperial, pink Electron, yellow King's Ransom and Sutter's Gold, apricot Tropicana, lavender Blue Moon, deep rose and lavender Paradise, cherry vanilla Double Delight, and the ivory blossoms tinged with pink of the legendary Peace. There are white floribunda Icebergs, a half dozen miniature roses, and two timeless species roses, the classic red Apothocary's Rose (*Rosa gallica officinalis*)

*Devotion gives each*
  *house a bough,*
*Or branch; each porch,*
  *each door, ere this,*
*An ark, a tabernacle is,*
*Made up of white-thorn*
  *neatly interwove;*
*As if here were those*
  *cooler shades of love.*
*Can such delights be in*
  *the street,*
*And open fields, and we*
  *not see't?*
*Come, we'll abroad; and*
  *let's obey*
*The proclamation made*
  *for May:*
*And sin no more, as we*
  *have done, by staying;*
*But, my Corinna, come,*
  *let's go a-maying.*
  ROBERT HERRICK,
  "Corinna's Going a
    Maying,"
  ll. 32–42 (1648)

and the pink-and-white-striped Rosa Mundi (*Rosa gallica versicolor*), which dates back to the early Renaissance, as well as other roses whose names I've yet to learn.

> *The longe day I shoop me*
> *for t'abide*
> *For nothing elles, and I*
> *shal nat lye.*
> *But for to loke upon the*
> *dayesie,*
> *That wel by reson men it*
> *calle may*
> *The "dayesye" or elles the*
> *"ye of day."*
>
> GEOFFREY CHAUCER,
> *The Legend of*
> *Good Women,*
> ll. 180–84 (c. 1386)

The roses in our gardens today date back thousands of years. No other flower has acquired so many legends, developed so much religious symbolism, or inspired so many poems. Rose cultivation began five thousand years ago in China and the Middle East. Roses, probably *Rosa gallica officinalis*, were described by the Greek poet Homer. The ancient Greeks cultivated roses, associating them with their love goddess, Aphrodite. They brought them to Rome, where they became the favorite flower of Venus, the Roman goddess of love. The Romans associated roses with joy, fertility, springtime, and mortality, strewing them at feasts and placing them on the tombs of the departed. Considering roses an aphrodisiac, the emperor Nero used thousands in his orgies, where roses perfumed the air and rose petals were stuffed into pillows for his amorous guests.

Early Catholic theologians rejected roses, associating them with lust and paganism, but by the late 300s, the church began changing its views. Saints Basil and Ambrose described the rose as the fairest of flowers, created without thorns in the Garden of Eden. By 400 C.E. the Catholic Church declared the white *Rosa alba* the symbol of the Virgin Mary, and the red *Rosa gallica* became associated with the blood of martyrs. In the early 500s St. Benedict planted a rose garden at his monastery in Subiaco, Italy. By 1321 Dante's *Divine Comedy* had described Heaven itself as a radiant celestial rose.[6]

Brought to England by the early Romans or later Crusaders, roses became an integral part of medieval life. They were carved into church screens and palace furniture, embroidered into tapestries, painted in manuscript illuminations, and immortalized in the classic *Romance of the Rose*. Roses were grown in monastery and church gardens and used to decorate altars and shrines. By the thirteenth century the medieval rose garden, or *rosarium*, had given its name to the rosary, the ritualized prayers said with beads and offered as a garland of roses to the Virgin Mary.[7]

Roses figured prominently in English history. In 1272 King Edward I planted a rose garden at the Tower of London and made the gold rose his symbol. During the War of the Roses in the 1400s, members of the House of York, descended from Richard II, chose the white *Rosa alba* as their symbol, while their Lancastrian cousins, descended from Henry IV, chose the red *Rosa gallica*. In 1485, when the Lancastrian Henry Tudor finally ended this war, claiming the throne as Henry VII and marrying Elizabeth of York, his symbol of national unity became the Tudor rose: a red rose with a white rose at its center.[8]

From the Middle Ages through the Renaissance, roses had many uses. Rose oil was used in medicines and to soothe chapped skin. Rose water was used as a cleanser, tonic, eyewash, and air freshener. People sipped rose syrup to calm the heart and cure fevers, coughs, and upset stomachs. They wove roses into garlands and made them into potpourri, for the scent of roses was believed to prevent or cure many diseases.

Roses were essential to medieval and Renaissance cuisine: enjoyed in rose sugar (dried rose petals mixed with sugar), served in tarts and preserves, and used to flavor honey, puddings, jellies, and wines. Men and women gave each other roses as expressions of love.[9]

Roses symbolized not only love but youth, beauty, and mortality. Their thorns recalled the inevitable pain of love, and their transience underscored life's mutability, a poignant reminder that even these perfect blossoms cannot last.

*Rose water and treakle, to comfort the hart*
*Cold herbes in hir garden for agues that burne,*
*That over strong heat to good temper may turne.*
THOMAS TUSSER,
"The Good Huswifelie Physiche"
(1580)

*I sent thee late a rosy wreath,*
*Not so much honoring thee*
*As giving it a hope, that there*
*It could not withered be.*
*But thou thereon didst only breathe,*
*And sent'st it back to me;*
*Since when it grows, and smells, I swear,*
*Not of itself but thee.*
BEN JONSON,
"Song, to Celia,"
ll. 9–16 (1616)

> *What is fairer then a*
> *  rose?*
> *What is sweeter? Yet it*
> *  purgeth.*
> *Purgings enmitie disclose,*
> *Enmitie forbearance*
> *  urgeth.*
>
> *If then all that worldlings*
> *  prize*
> *Be contracted to a rose;*
> *Sweetly there indeed it*
> *  lies,*
> *But it biteth in the close.*
>         GEORGE HERBERT,
>             "The Rose,"
>         from *The Temple*
>               (1633)

So much a part of their lives were roses that in 1620 the Pilgrims brought rosebushes with them to the New World, although America had its own native species. Roses have remained an American favorite. In 1986 the United States Congress declared the rose this country's national flower.

Yet the love of roses transcends national boundaries. In the seventeenth century Dutch horticulturalists bred new hybrid roses. In the mid-eighteenth century the predecessor of the floribunda rose was introduced to Europe from Japan. In the nineteenth century a new variety, the tea rose, was brought to England with a shipload of tea from Asia. More cultivation and breeding followed, producing La France, the first hybrid tea rose, in 1867.

My favorite hybrid tea rose, the legendary Peace, was developed in France in the late 1930s. As World War II began, samples of this rose with petals of golden ivory tinged with pink were sent to rose growers in Germany, Italy, and America. Robert Pyle, a Quaker in Philadelphia, grew the American sample. He introduced the rose at the Pacific Rose Society Exhibition in Pasadena, California, on April 29, 1945, naming it Peace. That same day the war in Europe ended, and Peace became one of the best-loved roses of all time.[10]

## GARDENING AS SPIRITUAL PRACTICE

THE GARDENS OF May remind me of a favorite Shakespeare sonnet:

> Shall I compare thee to a summer's day?
> Thou art more lovely and more temperate.
> Rough winds do shake the darling buds of May,
> And summer's lease hath all too short a date.[11]

For in Old England, May was the beginning of glorious summer: a time to enjoy the flowering of the natural world, contemplate its beauty, and celebrate its ephemeral loveliness. This month we too can take this tradition to heart.

## Maytime Celebrations

WHAT ARE YOUR earliest memories of May? Since I was a child I've claimed this month as my own, because my birthday is May 6. When my family lived in Maryland, my first-grade teacher told our class about May Day, helping us make simple baskets out of construction paper. I remember taking these small, cone-shaped baskets home, gathering wildflowers, then running around the neighborhood hanging May baskets on neighbors' doors.

This month I still feel like celebrating. One morning, when a woman friend called to say she'd gotten tenure, I put down my coffee cup and rushed outside to pick her a bouquet of roses, taking them to her office in a mason jar. The next week, as I left the house to meet a friend for lunch, I stopped on the way to pick her a bouquet of roses and lavender.

Throughout the year I make seasonal bouquets for our dinner table, bringing the beauty of the garden inside and watching the shifting flowers of the season. The roses of May grace our table with their special beauty and fragrance.

## Personal Exercise: Creating Your Own May Ritual

THE MONTH OF May invites us all to celebrate the loveliness of the season. This week look for ways to acknowledge your friends and fill your world with flowers. It takes only a few moments to turn a daily event into a Maytime celebration.

- As you walk through your garden, pick fresh flowers for your dinner table.
- Make a bouquet for a friend you're seeing that day.
- Bring roses or sprigs of lavender to brighten your desk at work.

🌿 If you don't have flowers growing in your garden, fresh bouquets are available at street stands or even in grocery stores this time of year.

Take time to enrich your life by honoring an old tradition: celebrating the fragrant blossoms and bountiful promise of May.

## Cultivating Relationships

ONE MORNING LAST week Bob walked into the living room and asked, "What's wrong with this plant?" The poor philodendron in the corner had visibly drooped. Its leaves were curled and the soil was bone dry. It was all too obvious that I'd forgotten to water it.

With the activities on campus and new growth in the garden, I've been taking my houseplants for granted. I've had this philodendron for years. My mother grew it from a cutting from the plant in her kitchen. The *Philodendron scandens oxycardium* is a patient plant, with vining, heart-shaped leaves, requiring only routine watering and occasional feeding. But now it was sitting in a corner of my living room, neglected and nearly forgotten.

I rushed out to the kitchen to get the watering can, feeling terribly guilty. Bob said he'd never forgotten to water his plants because he'd always had a routine. Every Saturday after cleaning his house, he'd water the plants. I'd done the same thing when I lived alone. But since we'd moved here I hadn't established a routine. As I watered the thirsty philodendron, I promised myself that every Sunday afternoon I'd water all the houseplants.

Later that day, as I checked on the philodendron, which looked a little better, my mind drifted back to when I was a young assistant professor. I'd joined a group for lunch on campus with a national consultant from Washington, D.C. A month or so later I was invited to consult with his team at a nearby university. We had a fascinating conversation over dinner and planned to meet again at our annual conference. In time fascination turned to romance, complicated by the distance between us. For a heady few months there were letters, long-distance phone calls, and meetings at conferences. Then the calls became less frequent, my letters went unanswered, and our contact dwindled to an occasional postcard. I began to wonder if I'd been imagining things: Were we friends, acquaintances, or something more? Weeks passed, then months, with no word at all. Eventually I started seeing someone else.

The next year, when my work took me to Washington, I met him for dinner. He greeted me with a smile, telling me about his new condominium, his travels and projects. When I told him about my new relationship, his face turned ashen. Visibly crushed, he said he'd just been busy; his feelings for me hadn't changed. "Relationships are living things, like plants," I told him. "People are not like books you can put on a shelf for months and expect them to be there when you want them. Relationships cannot live without care and cultivation." We left each other with an important lesson, too often overlooked in the rush of life.

The neglected philodendron has slowly recovered, but it was a valuable reminder. This week I called up an old friend and invited her to lunch. It had been months since we'd seen each other. We've both been busy, but if too much time goes by, friendships, like plants, will not survive.

## Questions to Consider

EVERY RELATIONSHIP NEEDS ongoing care and cultivation. Ask yourself these questions. If you answer yes to any of them, choose an action to help cultivate your relationship.

- Have I been out of touch with an old friend?
- Have I been out of touch with a favorite family member?
- Or—closer to home—do I need to cultivate my relationship with the one I love? Have our interactions been reduced to daily routine? What can we do to celebrate life and stay in touch on a regular basis?

Cultivation takes different forms for different relationships. If you live far away, a card, e-mail, or phone call can help put you back in touch. But paradoxically, as my philodendron has taught me, we often grow apart from those closest to us because we take them for granted. Many couples have developed special rituals to prevent this: a standing Saturday-night date, a regular vacation at a favorite romantic retreat, sharing walks around the neighborhood, picnics, or concerts in the park.

## Personal Exercise: Taking Action

WHAT IS ONE simple action you can take now to cultivate your relationship with someone you care about:

- 🐑 a phone call?
- 🐑 a card or e-mail?
- 🐑 a plan to get together?

However and wherever you choose, let the joyous month of May inspire you to cultivate new life in your relationships.

## Maytime Contemplations

CONTEMPLATING THE BEAUTY of nature has long been a part of spiritual practice, but it has often been overshadowed by devotion to work or otherworldliness. Western monasticism harbors opposing attitudes toward nature. The first, advocated by St. Augustine, urges people to condemn nature as fallen and sinful, avoiding it as a path of temptation. The second, advocated by the medieval Franciscan St. Bonaventura, praises nature as a pathway to God. In his *Itinerarium Mentis ad Deum* (The Mind's Road to God), Bonaventura encouraged people to use their five senses to contemplate nature. By opening their hearts to its beauty, they would discover the "traces," the lessons and insights left there by the divine Creator.[12]

The seventeenth-century poet Thomas Traherne expressed this Bonaventuran perspective when he wrote that

> You never enjoy the world aright, till you see how a sand exhibiteth the wisdom and power of God: And prize in everything the service which they do you, by manifesting His glory and goodness. . . . You never enjoy the world aright, till the Sea itself floweth in your veins, till you are clothed with the heavens, and crowned with the stars.[13]

Beauty—as gardeners we work for it, cultivating garden design, adding new bulbs or annuals, trimming, pruning, feeding. But the greatest beauty we often stumble upon unawares. While performing the routine tasks of watering and weeding, we discover unexpected blessings: new annuals that have reseeded themselves, a velvet-faced pansy blooming behind a mass of yellowing foliage, the first tiny green fruits on the tomatoes, the magic of sweet peas reaching out to embrace something close to them. As we look at the trees overhead, their sunlight-shadow patterns become a radiant green mosaic, an ineffable praise of life.

## The Healing Power of Beauty

THE HEALING POWER of beauty fills us with joy, sweeping away tension and stress. This simple lesson from medieval and Renaissance gardens can renew our spirits when we're chronically busy, when, exhausted by responsibilities, we see our lives as only sets of tasks to complete instead of an experience to enjoy.

## Personal Exercise: Breathing in Beauty

THIS MONTH FOLLOW the path of Traherne and give yourself the gift of "enjoying the world aright." Begin the next time you are in the garden, then practice this simple contemplative exercise at least once a day.

- Pause, take a deep breath, and look around you, slowly breathing out.
- Look for something beautiful: a tree, a flower, the sky overhead.
- Take a deep breath and breathe in its beauty. Then slowly breathe out.
- Smile and open your heart as you take another deep breath and release it.
- Then go about your normal activities.

Blending the garden around you and the garden within you, this simple but powerful exercise will leave you more relaxed, centered, and peaceful.

## GARDEN TASKS

IN MAY, AS our gardens turn from spring to summer, we face the alternating rhythms of endings and beginnings. When their spring bulbs die down to a mass of yellowing foliage, many gardeners are tempted to cut it all back to keep their gardens from looking messy. But living with a garden teaches us patience with a process we cannot rush. After blooming, spring bulbs take about six weeks to die back. During this period the leaves must be left on so the bulbs can store energy to bloom again next spring. The best we can do is deadhead the spent blossoms and let the remaining leaves do their work, cutting them back only when they die down completely. Meanwhile, we can satisfy our need to tidy up by spreading a layer of compost around the garden, which improves the appearance and texture of the soil while insulating it to retain valuable moisture during the hot summer months. We can also cultivate new beginnings, setting out fast-growing summer annuals to camouflage some of the yellowing foliage. In my garden the vigorous nasturtiums are doing a great job of this.

### Watering, Weeding, and Feeding

AS THE DAYS grow warmer I've been watering almost daily. The roses need lots of water, as do the tomatoes and the other plants growing in pots.

The warm May weather makes everything, including the weeds, grow well. To cut off weed seedlings before they get established, I got myself a new hoe and have been walking around the front gardens, cultivating the soil and decapitating the weeds, taking care not to disturb the herbs and flowers.

There are also lots of plants to feed. As the camellias, azaleas, and rhododendrons finish blooming, I've been giving them a special acid food and scattering some around the ferns growing nearby. I've fed the greens, tomatoes, and other vegetables weekly, and I just put new food stakes among the roses. The last ones went in in early April—I marked the date on my calendar. After six weeks it's time to feed them again. To keep the roses blooming until the end of the season, hammering in rose stakes at regular intervals is going to become a spring and summer ritual.

## Spring Trimming and Pruning

THIS MONTH THERE'S lots of trimming and pruning to do. Not only have I been deadheading spent roses and spring bulbs but I've also been cutting back the azaleas, camellias, and rhododendrons when they finish blooming. Some of their flowering branches were reaching across our front entrance and north path, making our garden into an obstacle course.

Pruning spring-blooming shrubs now will help shape them before they form new buds later this summer. This is also a good time to cut back old fern fronds. When the perennial candytuft in my front yard finished blooming, I trimmed all the dried flower heads with hedge shears, leaving a freshly groomed border of evergreen leaves. I also pinched back the chrysanthemums in the backyard to keep the plants from getting leggy so they'll produce better flowers this fall.

*Bankes newly quicksetted,*
*some weeding doo crave,*
*the kindlier nourishment*
*thereby to have.*
*Then after a shower to*
*weeding a snatch,*
*more easilie weede with*
*the roote to dispatch.*
THOMAS TUSSER,
*Five Hundred Points of*
*Good Husbandry*
(1580)

## Watching for Pests and Diseases

AS WE WALK around our gardens watering, weeding, and enjoying the beauty, we need to watch for pests and disease. Roses are susceptible to many diseases that attack their leaves: mildew, rust, and black spot. Not only are these infirmities ugly but they can be fatal. Through the process of photosynthesis, a plant's leaves use sunlight to transform carbon dioxide and water into vital nutrients. If its leaves are weakened, the plant can starve.[14]

At the first sign of mildew, rust (yellow-orange pustules on the undersides of leaves or yellow mottled leaves), or black spot (large, black spots on leaves), we need to take action. While watering the roses one morning I discovered black spot and rust on the inner leaves of some rosebushes (leaves without good air circulation are more vulnerable), so I cut off the most affected foliage. Then I walked around the garden like Lady Macbeth, determined to eliminate the spots, spraying the rosebushes with a sulfur-based fungicide.

Some of my friends give their roses a systemic fungicide as a preventive measure, but I don't want to put all that poison into the soil. Sulfur seems a milder response. After the rust and spots disappear, I'll spray the roses regularly with a nontoxic baking soda formula I got from a local Master Gardener, who swears it keeps her rose leaves green and healthy.

The warm weather and spring showers have been good for my kitchen garden. Tiny bok choy seedlings have sprouted in the south garden, and on May 5 the first two French green bean seedlings came up along the north garden fence.

Every morning I would run out to the back fence to see how many beans had sprouted. Two, four, six—the count was increasing day by day. But then the math seemed to be working in reverse—six, three, two. The plants were being destroyed as soon as they came up. In a week they were totally gone, and on the ground were the faint, shiny tracings of snails.

Once again our yard is filled with snails. It looks like this year we're having escargot in abundance, but no *haricots verts*. Even my "backup" Kentucky Wonder beans have been eaten.

Sometimes gardening means defending our boundaries. What *can* I do to keep the snails and slugs from eating all my vegetables and flowers? Last month I bought some rolls of copper foil to protect my pots of salad greens, and it seems to be working, so this month I put low wooden fences around the bean gardens and bordered them with copper foil. Following my friend Judy's advice, I sprinkled a line of salt outside the fences as an added barrier. Then I set more snail traps in the ground nearby and filled them with beer. After

### Environmentally Kind Spray for Roses

*Mix this simple solution, put it into a plastic spray bottle, and use it to prevent fungus, black spot, and mildew on roses or other garden plants.*

*1 gallon water*
*1 tablespoon baking soda*
*1 tablespoon light vegetable oil*

*Spray every ten days as needed. Do not spray in hot sun. Continued use will help your roses develop healthy, disease-free leaves.*[15]

that I planted more beans, lots of beans, two whole packages—enough for the snails and me.

## Planting Your Summer Garden

MAY BRINGS LOTS of gardening possibilities. Gardeners in zones 4 through 11 can plant summer annuals and many warm-weather vegetables, although some may need to watch for late frosts. Marigolds, zinnias, petunias, and sunflowers as well as many vegetables are available at local garden centers. Gardeners can plant cool-weather vegetables in zones 1 through 6 and warm weather vegetables in zones 4 through 11 (after the last frost in zones 4 through 6). In my region, we can sow beans, beets, carrots, corn, lettuce, radishes, spinach, and squash, and set out cucumbers, eggplants, peppers, and tomatoes, as well as annual herbs. I've just sown more lettuce seeds in planters inside our cool front courtyard, experimenting to see if I can grow salad greens in one part of the yard and tomatoes in the other to create fresh summer salads.

*Fine bazell sowe*
*In a pot to growe.*
THOMAS TUSSER,
*Five Hundred Points of*
*Good Husbandry*
(1580)

I've also set out some strawberry plants in another planter, which I surrounded with copper foil and placed in a sunny spot in our front courtyard. Strawberries grow wild in England and have been enjoyed there since Anglo-Saxon times. In the Middle Ages and Renaissance they were enjoyed raw, made into puddings, compotes, and jams, blended into elaborate sauces for meat, and used to soothe inflammations, comfort the spirits, and calm the heart. By the late Middle Ages these heart-shaped fruits were grown in the gardens of the aristocracy. King Edward III purchased strawberries for his garden in 1328, and in Shakespeare's *Richard III*, the bishop of Ely grew strawberries in his garden at Holborn. By the fifteenth century strawberries were sold on the streets of London.[16]

Strawberries have always been my favorite summer fruit. As I child I enjoyed strawberry shortcake and strawberry ice cream. I even planted strawberries in our yard, but I was a frustrated gardener, because the earth kept shifting beneath my feet. My father was an Air Force colonel, and it seemed whenever I'd plant a patch of strawberries, we'd get transferred. When I was twelve I planted strawberries outside our house in Grandview, Missouri. We were transferred that year to Eglin Air Force Base in Florida. Then my father was sent to command a radar site outside Omaha.

The radar site was in Nebraska farm country, and there were only six houses on the base. There were few children, none my age. Each day my brother and I were driven to school in Omaha in an Air Force staff car.

Isolated from my peers, I spent the next year and a half doing lots of reading. I also learned to cook, subjecting my family to culinary experiments at least once a week. In the spring I cleared a hundred-square-foot plot of ground behind our house. After pulling up all the wild grasses and turning over the soil, I planted strawberries. I was out working in my field one day when my father came home with the news—we were being transferred to another radar site, in Olathe, Kansas. My heart sank. I looked down at the strawberries I would never harvest, then slowly walked into the house to begin packing.

> Camomile, the more it is trodden on,
> the faster it grows.
> WILLIAM SHAKESPEARE,
> 1 Henry IV,
> II.5, ll. 365–66 (1598)

Today I prefer to grow strawberries in containers, perhaps because of childhood conditioning but also because they're less susceptible to snails and slugs. My strawberries are already growing well, benefiting from the warm sun, and I look forward to abundant summer harvests.

These warm days of May are also a good time to sow or set out herbs. This month I've planted more basil and dill seeds in pots surrounded by copper foil. Since I enjoy chamomile tea, I decided to grow some chamomile as well. At the local health food store I bought a packet of organic heirloom chamomile seeds. Unlike the new hybrid varieties, heirloom seeds have been passed down for over a hundred years. With all the challenges to our ecosystem brought about by chemical farming and genetic engineering, it's comforting to know that some seeds, some plants, have withstood the test of time.[17]

The seeds I bought and planted are German chamomile (*Matricaria recutita*), which grows about two feet tall. Another variety, Roman or English chamomile (*Anthemis nobilis* L.), is a ground cover, four to six inches high, used in medieval England for strewing, potpourri, and many herbal remedies. Both varieties of chamomile have been made into tea, used to cure indigestion, calm the nerves, and, when taken with honey, soothe sore throats. A good companion plant, chamomile improves the health of plants growing nearby. The tea also has antiseptic qualities. It helps heal minor skin inflammations and, when sprayed on seedlings, prevents damping off, a fungus disease that kills seedlings by rotting their stems at the soil surface.

### Making Chamomile Tea

*To make your own chamomile tea, pick the flowers and dry them in a cool, dark place with plenty of air circulation. When the flowers are thoroughly dry, store them in a jar as you would any other herbs. For tea steep a heaping teaspoon of dried chamomile flowers in a cup of boiling water for 5 minutes. Then strain to remove the flowers, and serve with honey to taste.*

## GARDEN REFLECTION

### *Cultivating Order in Our Gardens and Our Lives*

YESTERDAY I SPENT most of my time in the garden cleaning up: mowing the small plot of grass in our front yard, trimming around the cobblestone walk, picking up fallen magnolia leaves from our brick courtyard, deadheading the roses, and digging up a few errant dandelions. So much of gardening revolves around the routine tasks of maintaining order. Not as dramatic as setting out new vegetables or colorful annuals, these tasks are nonetheless essential to maintaining our gardens' health, and perhaps our own.

The postmodern world rushes us from one event to the next, assaulting our senses, raising our adrenaline levels, and bombarding us with so much information that we easily lose touch with the natural rhythms of life. Many of us become bored with routine tasks, seeing them only as chores to be rushed through or dispensed with entirely.

Many of my friends hire gardeners to do their routine mowing, trimming, and weeding so they can spend more time at work. Perhaps this choice is right for them. Faced with multiple demands on our time, each of us must determine what to cultivate, what to prune from our lives. We can delegate a number of tasks these days. Computer companies have arranged for laundry services, meals, and car maintenance on site so their employees can work longer hours. Those of us with Internet connections hardly need to leave our desks. We can purchase consumer goods on-line. Even groceries can be ordered from a Web site and delivered to our doors by appointment.

The tools of our technology can bring greater convenience or greater confusion to our lives. The critical factor is balance. Although delegating some tasks keeps our lives from becoming overcrowded, if we delegate too much we become spiritually impoverished. A challenging job may bring us an adrenaline rush and fill our lives with drama, but we cannot remain healthy in a state of continuous arousal. As much as we need challenge to give our lives meaning, we also need the nurturing rhythms of routine, rituals of order to restore our balance, return us to ourselves, and ground us in ordinary time.

When we become too busy, these very rituals often fall by the wayside. Whenever I'm out of balance, juggling too many roles and responsibilities, I begin forgetting things. May is the busiest month on campus. As the academic year draws to a close, everyone rushes to finish the year's projects, stacking more meetings, reports, and deadlines on top of a full load of classes and research, then adding a final layer of official ceremonies, receptions, and evening events. Every May I plow through the endless round of activities with what Thoreau called "quiet desperation." My schedule becomes too cluttered, my mind too distracted, to be fully present to the people in my life or to this lovely month when spring blossoms into summer. Last week I could feel my schedule closing in on me as I started forgetting things again. I "lost" my prescription sunglasses, then found them five days later in my jacket pocket. I left my silver pen on my desk at work and "found" it again on Monday. Last Friday, after spending a lovely afternoon having tea with Tracey and Chris, I left my box of birthday gifts on a table in Chris's study as I dashed home to finish another report.

At home I realized my mistake, called Chris, then went out to the garden. Feeling the earth beneath my feet for the first time in days, I slowly walked around deadheading roses and picking up a few fallen magnolia leaves. This week I've pruned back my overgrown schedule, spent more time in the garden, and stopped forgetting things. For me the simple tasks of garden maintenance are always rituals of order. Time in the garden forces me to slow down, get back in touch with myself, and see my professional life from a healthier perspective.

My garden has taught me the wisdom of alternating rhythms of action and contemplation, a tradition that goes back over fifteen hundred years, when the Rule of St. Benedict established a model of life for Western monasticism that balanced work and worship: *laborare et orare*. With their days structured by periods of prayer, study, and manual labor, medieval Benedictines combined the work of hands and hearts in alternating rhythms of devotion. Whether a monk's duties involved weeding the garden, baking bread, cleaning the refectory, or graciously welcoming a guest, every task was balanced by its opposite and sanctified by loving attention.[18]

The Benedictine balance of action and contemplation can help us find our own paths to wholeness. More than a series of mindless chores, the routine tasks of gardening release us from the staccato pace of the world around us, returning us to more natural rhythms. Simple actions such as staking the tomatoes, picking up and composting fallen leaves, watering, weeding, and trimming can become valuable rituals in the patterns of our lives, keeping the gardens within and around us in good order.

# SUMMER

# Summer Garden Checklist

## EARLY SUMMER

**Plant:** Sow or set out summer annuals, herbs, and warm-weather vegetables. Plant new perennials and container-grown trees and shrubs if the weather is not too hot. Finish planting summer bulbs.

**Water:** Water regularly if it does not rain often in your region. The best times to water are early mornings or late afternoons. Soak the roots of roses and avoid spraying their leaves, which encourages mildew. Pay special attention to new plantings. Container-grown plants will need watering daily, during hot spells twice a day. Check sprinklers and drip systems for proper functioning.

**Weed:** Keep dandelions and other weeds pulled so they don't rob your garden plants of food and water.

**Feed:** Feed roses, summer annuals, and vegetables. Container-grown vegetables need feeding more often. Check the fertilizer package for specific instructions. Feed spring-blooming shrubs, citrus and fruit trees.

**Cut back:** Prune frost-damaged trees and shrubs. Trim spring-flowering plants, vines, and shrubs after they bloom. Prune wandering wisteria vines. Deadhead flowers from spring bulbs but leave foliage in place. Pinch back the ends of chrysanthemums. Deadhead roses to encourage blooming. Thin vegetable seedlings. Pinch suckers off tomatoes. Thin young fruit from heavily laden fruit trees.

**Cultivate:** Put mulch around roses, azaleas, camellias, and rhododendrons to retain moisture and insulate the soil from summer heat. Support climbing plants. Stake tomatoes. Put straw around strawberries.

**Harvest:** Harvest fresh fruits, herbs, and vegetables.

**Watch for:** Check for pests and mildew, especially on roses. Spray aphids with insecticidal soap and mildew with fungicide. Set out traps or bait for snails and slugs. Watch tomatoes for green hornworms. Hand-pick them with gloves and discard.

**Enjoy:** Enjoy the roses of summer, abundant harvests, and watching your garden grow.

## MIDSUMMER

✖ *Plant:* Set out annuals for bright summer color as well as herbs and some summer vegetables. In cooler regions sow fall vegetables (check seed packets for correct planting time for your area). Order spring bulbs. Set out container-grown trees and shrubs when it is not too hot.

✖ *Water:* Water regularly if it does not rain often in your region, giving special attention to newly planted and container-grown plants.

✖ *Weed:* Keep dandelions and other weeds under control.

✖ *Feed:* Continue to feed roses, annuals, vegetables, flowering shrubs, and fruit trees.

✖ *Cut back:* Remove spent blossoms from flowering shrubs and trim them back after they bloom. Prune wandering wisteria vines, bushes, and hedges. Pinch back the ends of chrysanthemums. Deadhead roses regularly. Trim spent blossoms from annuals to keep them blooming. Thin vegetables as needed. Pinch suckers off tomatoes.

✖ *Cultivate:* Mulch around roses, trees, and shrubs to retain moisture and insulate the soil. Stake gladioluses, dahlias, tomatoes, and climbing vegetables. Cultivate the soil, adding compost. Trim leaves from spring bulbs when they turn yellow and dry up. Begin digging up, dividing, and replanting irises.

✖ *Harvest:* Harvest herbs and keep them trimmed throughout the season. Harvest ripe fruits and vegetables.

✖ *Watch for:* Continue to check for pests and disease. Treat when necessary. Watch tomatoes for green hornworms, pick them off, and discard.

✖ *Enjoy:* Enjoy outdoor meals with fruits and vegetables from your garden.

## LATE SUMMER

✖ *Plant:* Start ordering and planting spring-flowering bulbs. Plant fall vegetables (in warmer regions, you may need to sow your seeds indoors). Plant evergreens if the weather is not too hot.

✖ *Water:* Water regularly if it does not rain often in your region, giving special attention to newly planted and container-grown plants, newly sown seeds, and seedlings.

✖ *Weed:* Keep pulling and hoeing up weeds.

✂ *Feed:* Feed roses for the last time six weeks before the first frost. Continue to feed vegetables, annuals, and summer-blooming shrubs. Feed chrysanthemums three times between now and mid-September for abundant autumn flowering. Feed azaleas, camellias, rhododendrons, and other spring-flowering shrubs for the last time this year.

✂ *Cut back:* Remove spent blooms and cut back flowering shrubs after they finish blooming. Trim wandering wisteria vines, bushes, and hedges. Deadhead roses, dahlias, and summer annuals to keep them blooming. If your strawberries have stopped bearing, trim off old leaves and unwanted runners. Take out unhealthy plants, remove the old straw, and cultivate the soil between rows.

✂ *Cultivate:* Stake gladioluses, dahlias, and tomatoes. Support heavily laden branches of tomatoes and fruit trees. Prepare the soil for fall vegetables. Protect cool-weather vegetables with shade. Continue dividing and replanting irises, completing the task four to six weeks before the first frost. Mulch around plants to conserve moisture. In cooler regions root cuttings of herbs and geraniums to grow through the winter in pots indoors.

✂ *Harvest:* Harvest tomatoes and other vegetables. Trim herbs to enjoy in summer meals and dry some to put away for winter.

✂ *Watch for:* Check and treat for pests and disease. Pick up fallen fruit from the ground to prevent disease.

✂ *Enjoy:* Enjoy all the flowers of late summer and easy outdoor meals with summer vegetables.

# SIX

## *June: Presence, Patience, and Planting New Seeds*

Gather ye Rosebuds while ye may,
Old Time is still a-flying:
And this same flower that smiles today,
To morrow will be dying.

The glorious Lamp of Heaven, the Sun,
The higher he's a getting;
The sooner will his Race be run,
And neerer he's to Setting.

ROBERT HERRICK,
"To the Virgins, to make much of Time,"
ll. 1–8(1648)

YESTERDAY AS I was deadheading the roses, that line from Herrick's poem, "Gather ye Rosebuds while ye may," danced through my head. Looking around for rosebuds, I cut some in four delicious colors, yellow, tangerine, rose, and melon, to bring inside. In June's 90-degree heat, roses bloom and fade fast. What I don't pick to enjoy today I will only be deadheading tomorrow. As Herrick's poem reminds us, the rose's blooming cycle and time itself move inexorably along. The art of living, like the art of gardening, is much more than what we do, whether it's planting, watering, trimming, or pulling weeds. In the gardens of our lives, beneath it all is the lesson of presence, the art of enjoying the present moment.

Presence can take many forms, from a lively *carpe diem* (literally, "seize the day") enjoyment of life to the quiet lessons we learn in our gardens. In June the promise of summer stretches before us. The days are filled with light, growing progressively longer until the summer solstice on June 21, the longest day of the year. In early Britain before the advent of Christianity, people feasted outdoors and lit bonfires at this time of year, celebrating Midsummer Day on June 24. The custom of outdoor feasting and games endured throughout the Middle Ages. In Scandinavian countries Midsummer is still a major holiday, celebrated outdoors with feasting, drinking, and good fellowship. Even in America many people unknowingly repeat these early rituals, lingering in their gardens on a summer evening, lighting their barbecues, enjoying outdoor meals with families and friends.

The medieval Catholic Church substituted its own holy days for earlier pagan rituals. June celebrations began the week after Whitsuntide with Trinity Sunday's colorful church processions, feasting, and dancing. The following Thursday, the Feast of Corpus Christi, was celebrated with elaborate mystery plays in which village guilds acted out biblical scenes. On June 24 the church celebrated St. John's Day, marking the birth of John the Baptist.[1]

## GARDEN GROWTH

WHILE THE ROSES are in full bloom, the rest of my June garden is still in transition from spring to summer, with only a few clumps of fading foliage to recall the glories of spring. But if we look carefully at our gardens, we can find new flowers appearing as each season moves on. The heavenly bamboo (*Nandina domestica*) bush growing outside my study window has sprigs of small white flowers. I never saw this kind of plant blossom before.

The herb garden looks a little bare, bereft of its abundant spring bulbs.

But its floral tapestry has changed only in scale. There are tiny violet flowers blooming on the thyme, the summer dahlias have put forth their first shoots, and the chamomile has come up, its delicate foliage like green lace. Many annuals have reseeded themselves, creating new variations in the tapestry of herbs, red and gold nasturtiums, white alyssum, and blue lobelia.

On the fence between our houses, Rhonda's sweet peas (*Lathyrus odoratus*) are filled with fragrant blossoms of rose, scarlet, magenta, and violet. In the friendship garden groves of spiky liatris (*Liatris spicata*) raise their violet plumes in the summer breeze. The first pale purple blossoms have appeared on the dahlias, and two gladioluses (*Gladiolus grandiflora*) are blooming in shades of deep rose. The lilies of the Nile (*Agapanthus orientalis*) raise their buds to the sun, and the crocosmia (*Crocosmia masonorum*) have put out long stems filled with copper-colored buds, preparing to burst into bloom in the warm days ahead.

Along the north side of the house and around the deck in back, the hydrangeas (*Hydrangea macrophylla*) are blooming in clusters of white, rose, lilac, mauve, and pale blue. The sunlight filters through the trees onto their blossoms like the play of light in French Impressionist painting.

> *A gardyn saw I ful of*
>   *blosmy bowes*
> *Upon a ryver, in a grene*
>   *mede,*
> *There as swetnesse*
>   *everemore inow is,*
> *With flowres white, blewe,*
>   *yelwe, and rede.*
>   GEOFFREY CHAUCER,
>   *The Parliament of Fowls,*
>     ll. 183–86 (c. 1380)

Inside the front courtyard it's hard to miss the new bush blooming in the corner. With its bright yellow blossoms abundant as summer sunshine, this plant is St. John's wort (*Hypericum perforatum*). Early Christian priests named it for John the Baptist because it blooms around his feast day, June 24. In medieval times its leaves were eaten in salads and made into an oil for healing wounds. Sprigs of the plant were hung in doorways to drive away witches and evil spirits.

St. John's wort was once considered an invasive weed in the American West. Known as Klamath weed in the 1950s, this plant spread across more than 2 million acres of open land in Northern California and Southern Oregon. Because eating it made their cattle's skin sun-sensitive, ranchers

*If you live in an area with mild winters, you might want to plant St. John's wort in your garden. It's a hardy shrub that requires little care and grows well even in poor soil. In fact, it's so vigorous it can become invasive. It provides a display of yellow blossoms for at least two months from late spring to midsummer. There are more than a dozen varieties of this perennial. Some grow up to four feet high, while others, six to twelve inches high, make good ground cover.*

controlled the weed with predators and herbicides, eliminating 90 percent of it by the 1960s.[2] Now, of course, St. John's wort is considered a valuable herb, harvested for its antidepressant qualities. There's a lesson here about not being too quick to dismiss something natural as worthless, because in time we may find it extremely valuable.

In my vegetable gardens the salad greens and bok choy are growing well, the basil, dill, and chamomile I sowed last month have become vigorous young plants, and the strawberries are filled with blossoms and tiny green berries. Early this month I discovered the first tomatoes of the season, bright green and smaller than marbles. By Midsummer the tomato vines were covered with green fruit. Each day as I watered them I found myself increasingly impatient, wondering when they would ripen. Gardening often reveals what we need to cultivate within us. In my case the perennial lesson is patience: patience with nature's process, patience in achieving my goals.

## GARDENING AS SPIRITUAL PRACTICE

### Planting New Seeds

THE ORIENTAL POPPY seeds I planted last fall never did come up. By June I'd thought they'd be adding their vibrant color to the iris garden. I don't know what happened, but instead of promising green shoots, each day as I watered and watched I saw only faded foliage and bare spots where once spring bulbs had grown.

It was time to change my plans for my summer garden tapestry. This week I walked around the front gardens scattering the seeds of violet and white sweet alyssum (*Lobularia maritima*) in all the bare spots. Sweet alyssum will germinate in eight to fifteen days and be blooming within six weeks, enhancing the garden with lacy blossoms of violet and white.

If some of the seeds you planted did not come up, you still have time to sow more or set out new bedding plants. Choose a variety that will grow well in your area, and you'll soon have new possibilities blossoming before your eyes.

I found this lesson repeated on another level when I met with my student friend Anna, who is applying to graduate school this fall. In many ways planning our careers is like planting seeds. We have to select the program or job we desire, do the necessary cultivation, cast our seeds and hopes out into the world, then wait to see what happens.

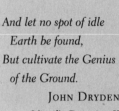

*And let no spot of idle Earth be found, But cultivate the Genius of the Ground.*
JOHN DRYDEN,
*Virgil's Georgics*, II,
ll. 49–50 (1697)

Not all our seeds will germinate—not all applications will succeed. I told Anna I'd applied to at least eight schools. If some didn't work out, I wanted to have others to turn to. "It's like planting seeds," I told her.

Anna smiled in recognition. "That's right," she said. "I planted two packets of seeds. One didn't amount to anything, but the others have just come up."

## Questions to Consider

THIS MONTH IT'S time to consider the seeds you've been planting in your life.

- How well have your seeds, your recent plans, succeeded?
- Have they made a positive contribution to your life's design?
- If not, can you think of other possibilities, other seeds you can plant?
- Brainstorm: Write down as many actions as you can think of, or meet with a friend to generate new possibilities.
- Choose some possibilities you'd like to pursue and make a list. These are your new seeds.
- Choose one seed and take action on it this week.

### Patience in the Middle Season

I LOVE BEGINNINGS and endings: planting the seeds of new possibilities, starting new projects, completing them, and celebrating the joy of harvest. But as I realized this month with my tomatoes, the in-between stage, the long middle season, tests my patience. At times progress is so slow nothing seems to be happening. I feel frustrated and ineffective, and often try to push the process.

When I was in graduate school my friend Pat had some beautiful avocado plants she'd grown from seeds. So I planted an avocado seed in a pot and waited. Weeks went by, but nothing happened. I finally dug up the seed to see what was going on. The seedling was reaching to the surface, about to emerge from the soil, but my impatience had killed it.

The month of June is filled with garden growth. Yet for many of us this is a slow middle season, still too early for harvest. The fruits of summer will come later, and the most abundant harvests only after summer has passed. The lessons this month are presence and patience: to be patient with nature's timing and present with the possibilities in our midst.

Cultivating patience is a special challenge for highly motivated people intent on reaching our goals. But patience with process is essential to peace of mind. Most of life is spent in the middle season, between the excitement of new beginnings and the fulfillment of conclusions. Our projects, careers, and relationships all have long summer seasons. Our enthusiasm for new projects often wanes, the glamorous new career becomes daily routine, courtship settles down into the everydayness of living with the one we love.

If we focus too much on future goals, we can miss the rich opportunities before our eyes. During academic advising at the university, most students concentrate on getting required classes "out of the way" so they can graduate. But this spring many graduating seniors said they wished they'd spent more time enjoying their college education, because it had all gone by so quickly.

When we find our minds racing ahead of us in the middle season, we can cultivate patience by consciously slowing down, taking a deep breath, and bringing ourselves back to the present moment. During training in aikido, a martial art that is also a spiritual exercise, my *sensei* (teacher), Sunny Skys, reminds our class to stop rushing and take a deep breath to regain our centers. When aikido students lack confidence, they often push through a technique, trying to get it over with quickly. Yet the point is not to get the technique over with but rather to open our hearts, embrace the process, and learn from it.

"You don't dance to get to the other side of the room," Sunny says. In aikido presence improves our performance. Slowing down makes us concentrate on the energies within and around us, becoming more aware of nuances, more open to new insights. In every process of our lives, greater wisdom is ours when we embrace the middle season.

## Personal Exercise: Embracing the Process

- Do you find yourself in a middle season with a project, career, educational process, or relationship?
- Have you responded with impatience: rushing, complaining, trying to push things?
- If so, it's time to reframe the process, concentrating not so much on getting it over with as experiencing what it has to offer.
- Take a few minutes to visualize yourself slowing down, enjoying the dance, opening yourself to the wisdom you find.
- Take a deep breath and release it, feeling yourself fully present, fully here, right now.
- Smile as you take another deep breath, embrace this new vision, and accept it into your life.

## The Power of Growth

YESTERDAY I NOTICED new roots on the rosemary I'd placed in a jar of water in my kitchen window. In a few more days I'll put it in some potting soil and let it grow into a new plant.

This time of year the power of growth is so evident. Cuttings take root, seeds germinate, and plants grow so quickly. The dill I planted last month

**Rooting Herbs and Flowers**

*You can grow many plants from softwood cuttings. Some of the easiest are herbs: rosemary, basil, mint, oregano, thyme. Cut off a healthy stem, three to five inches long, just below a leaf node. Pinch off the bottom leaves. Place the cutting in a container of water or moist vermiculite so that half of it extends above the growing medium. Set it in a window with filtered light. When your cutting develops roots (in four to six weeks), you can plant it in the garden or in a small pot, keeping its roots moist to help it through the transition.*

is already three feet tall. The medieval abbess and holistic healer St. Hildegard of Bingen called this greening power of life *veriditas*, evidence of God's grace in the world, which she saw as the key to vitality, goodness, and health.[3]

This month, as you witness the greening power of life in the garden around you, ask yourself, "What new growth am I seeing in *my* life?" Is there anything in your life you'd like to increase? Some healthy new habit you'd like to take root? Let these long days of summer, filled with warmth and light, inspire you to reach out, grow, and increase your *veriditas*, your joy and vitality.

### Choosing What You Want to Grow

SOMETHING IS ALWAYS growing in our gardens, whether we cultivate it or not. So it is with our time. Our days can too easily become cluttered with demands and activities put there by others, like invasive weeds choking out the space we need for growth. My friend Ed Kleinschmidt Mayes, who founded the Creative Writing Program at Santa Clara University, asks his students to set aside at least seven hours a week to write. This requires discipline: blocking out the time, fencing it in, guarding it from distractions. An accomplished poet, he sets them a good example, letting them know when he'll be available for them and when he's scheduled his own writing time. By mastering this lesson his students have excelled, winning poetry prizes and scholarships, reaching out to achieve their dreams.

In our lives as in our gardens, if we don't set aside the time to cultivate something we care about, it will not grow. Each day the world offers us many opportunities and interruptions. We can fill our calendars so full we barely have time to breathe, let alone live creatively. As vigilant gardeners we need to recognize and eliminate weeds and their counterparts from our lives.

What do we choose to grow in our gardens? The shoots of invasive grasses? Dandelion seeds blowing in from a neighbor's yard? Or our own herbs and flowers? If "eternal vigilance is the price of liberty,"[4] it is also the price of living an authentic life.

### Personal Exercise: Beginning a Creative Practice

IS THERE A creative path you'd like to pursue: writing, music, painting, or photography? An exercise or meditative practice? Some other path of personal growth?

- If so, resolve to cultivate a plot of time to devote to this art.
- Schedule at least two hours this week.
- Block out your time on your calendar.
- Next week try for three hours and gradually work up to seven.
- At the end of the month, ask yourself how your new practice makes you feel. What new insights and opportunities have you harvested?

### The Invisible Work Is the Most Important

WHILE OUR DOMINANT culture emphasizes surface accomplishments, in our inner gardens the most important work is often invisible, the work that no one sees, no one knows but you. It's the promise you keep to yourself, the discipline it takes to persevere in something you value.

Do you have a personal practice that cultivates your spirit, strengthens your character? My practice is aikido, a nonviolent martial art based on ancient samurai exercises, a way of resolving conflict within and around us by harmonizing opposing energies. What aikido *looks* like is a combination of judo and ballet. What aikido *feels* like, when done well, is smoothness, strength, and flow, a joyous feeling of moving from center, tranforming an attack into a spiral of harmony, blending with the energy of the universe. When I'm off center, aikido feels like awkwardness, embarrassment, and ego. Either way it is a powerful mind-body therapy, helping me tune into myself on deeper spiritual and kinesthetic levels, strengthening my body, mind, and spirit.

Two of my friends, Sunny Skys and Miki Yoneda, are aikido black belts. Their dojo, or training hall, is in Fremont, about thirty miles away. So two evenings a week I leave the house around 6:00 P.M. and spend up to an hour in rush-hour traffic getting to the dojo, where I put on my training uniform, my gi and hakama, and bow in to begin class. Then I spend an hour or more on the pale green mat, taking falls, throwing my partners, learning new techniques, and gaining new insights into old ones. I return home around 9:00 P.M.

What do I gain? Balance, respect, a reverence for life, a greater understanding of myself and my place in the universe. I'm always glad when I get to the dojo, but it takes discipline to keep up my training. It's all too easy to become "too tired" or "too busy" to go. When I arrive in Fremont and bow onto the mat, I have already completed an important part of my practice. The most important work is invisible: the discipline it takes to get me there.

## Questions to Consider

- What is your personal practice?
- What invisible work do you need to get you there?

## GARDEN TASKS

### Watering

THE LAST PACKET of green beans I planted never sprouted, so I tried to figure out why. First I thought it was the soil. Was the ground in the north garden too hard? But no, I had cultivated it, adding compost and potting soil. If not the soil perhaps it was the weather. In May our weather changes abruptly from spring to summer, with showers giving way to warm, sunny days. At the crucial germination time, when seeds need to be kept moist, it may have been too dry for the beans to sprout.

An essential task in June is making sure your garden gets enough water. Germinating seeds and newly transplanted plants are especially vulnerable to drought. Before planting any more beans, I bought a sprinkler hose and ran it all the way down the north side of the house, through the green bean beds on both sides of the fence.

### Being Water Wise

THIS SUMMER, REMEMBER to be water wise. Most plants need at least one inch of water a week (either rain or irrigation). Here are some helpful reminders:

• *Deep watering is essential.* Shallow watering evaporates quickly and may do more harm than good, encouraging plants' roots to grow near the surface, where they can be scorched or dry out.
• *Try not to get the plants' leaves wet.* Water drops can focus the sun's rays like magnifying glasses, burning plants' leaves. Damp leaves can also cause mildew on roses and many vegetables.
• *Use a soaker hose or drip irrigation* to deliver water to the plants' roots.
• *Or make basins around your plants* by scooping the soil from around their stems. Fill the basins with water and let it soak in.
• *Water in the early morning or late afternoon,* when it's cooler and the water won't evaporate.
• *Improve your soil's water retention* by enriching sandy soil with organic matter.
• *Spread a two- to three-inch layer of mulch* on your soil to preserve moisture.

New bedding plants in your flower garden need extra watering until they get established, then regular watering throughout the season. Roses need to be deep-watered with at least one gallon per bush. In hot weather they may need as much as five gallons a day. Other plants need more water if any of the following are true:

- *You've just transplanted them.*
- *Their seeds are germinating.* Keep the soil moist and don't let the seeds dry out.
- *They're tiny seedlings.*
- *They're setting buds.* (This includes many spring flowers and vegetables as well as perennials such as camellias and rhododendrons, which set buds in late summer.)
- *You've just fertilized them.*
- *They're flowering and bearing fruit.*
- *They're leafy vegetables* (such as spinach or lettuce).
- *They're root vegetables* (such as carrots, turnips, beets).
- *They're vegetables such as peas, beans, and corn,* which need abundant water throughout the growing season.
- *They're growing in containers.* Container plants can easily dry out in hot weather and may need watering as much as twice a day.
- *They're growing in the hot sun, in windy areas, beside walls, in sandy soil, or on steep slopes,* where water rapidly evaporates or drains away.

The container plants inside our front courtyard are sheltered from the hot afternoon sun by the Japanese maple (*Acer palmatum*). But to make sure they get enough water, I've set a beautiful green-and-gold watering can beside them. Once a day I check the plants' moisture level and sprinkle them with water.

## Choosing Drought-Resistant Plants for Dry Areas

IT'S WISER TO harmonize with local conditions than to fight them. If you live in a temperate region and areas of your garden are prone to drought, you can choose plants suited to a Mediterranean climate. Many herbs, trees, and shrubs thrive in dry, sunny conditions. Your local garden center can recommend varieties especially suited to your area, but here are some possibilities:

- **Trees**: eucalyptus, crape myrtle (*Lagerstroemia indica*), sweet bay (*Laurus nobilis*), olive (*Olea europaea*), oak (*Quercus*)
- **Shrubs**: bottlebrush (*Callistemon*), broom (*Genista*), heavenly bamboo (*Nandina domestica*), oleander (*Nerium oleander*), plumbago (*Plumbago auriculata*)
- **Herbs**: lavender (*Lavendula*), rosemary (*Rosmarinus officinalis*), sage (*Salvia*), thyme (*Thymus*)
- **Flowers**: blanket flower (*Gaillardia*), sweet alyssum (*Lobularia*), California poppy (*Eschscholzia californica*), lamb's ears (*Stachys byzantina*), fortnight lily (*Dietes*), lilies of the Nile (*Agapanthus*)

## Weeding and Trimming

SINCE WEEDS SEEM to grow overnight in the summer heat, June is a time for vigilant action. I spend a few hours each week on weed patrol, walking around the yard with my garden hat to shield me from the sun, my tool belt, hoe, and a large bushel basket. Staking out one area at a time, preferably after watering, when the ground is soft and easier to work, I look for weeds, cutting off the small ones with my hoe or pulling the larger ones with the knife from my tool belt.

The herb garden was looking a bit disheveled, but a day of pulling small weeds and grass shoots has made a major difference. Now the herbs and alyssum look more attractive, and I even noticed some small volunteer lobelia. Strange how removing what we *don't* want helps us discover what we do. Clutter, even in a garden, creates mess and confusion, preventing us from seeing the beauty hiding in plain sight.

---

*Checking for Soil Moisture*

*Check to see if your plants are getting enough water by literally getting in touch with the earth. Poke your finger into the ground. If the surface is dry but the ground is still moist one or two inches beneath the surface, your plants' roots have enough water. If the ground is dry beneath the top layer of soil, you need to water your plants—now. Chronic water shortage will stunt plants' growth. Severe drought can kill plants.*

Another task this month is removing spent blossoms and trimming spring-flowering shrubs after they bloom. The backyard needed lots of attention. The tea tree (*Leptospermum scoparium*) and heavenly bamboo (*Nandina domestica*) outside my study had grown so high they were beginning to block the window. I waited until their blossoms had faded, then gave them both a good trimming with the hedge shears. Nearby the wisteria (*Wisteria sinensis*) was putting out long, unruly tendrils and reaching into the liquidambar tree (*Liquidambar styraciflua*), so I took some long-handled shears and trimmed it back as well, then cut back some ferns that were growing over the benches beneath the wiseria arbor. Our yard now looks much more inviting—in time for outdoor gatherings this summer.

*Fie on't, ah fie, fie! 'Tis an unweeded garden That grows to seed; things rank and gross in nature Possess it merely.*

WILLIAM SHAKESPEARE, *Hamlet*, I.2, ll. 135–37 (1603)

## Staking and Mulching

THIS WEEK I bought four bags of organic compost to mulch the ground in the front garden. The temperature has been in the nineties. The sun beats down, baking the bare earth and the plants' roots underneath. Mulching with compost keeps the moisture in, protects the plants, conserves water, discourages weeds, and enriches the soil.

As they grow taller your flowers and vegetables may need tying up and staking. I put bamboo stakes beside two gladioluses this week and bought steel spirals for the tomatoes in the side garden. Early Girl was falling all over herself, the cherry tomatoes were trying to climb the fence, and the Sweet 100s were outgrowing their cages. These spirals will give them added support.

## Watching for Pests and Disease

AS ATTRACTIVE AS our gardens are to us, they also attract pests. For months the snails and slugs have been eating the tender growth of many new flowers and vegetables in my garden. The traps I've set out have been collecting dozens of snails and slugs each day in their irresistible pools of beer. Another way to combat snails is to seek and destroy their hiding places: fallen leaves and damp, heavy ground foliage. I pulled out the wildflowers and grass growing by the large rock in the herb garden, a popular snail hideout. I also picked up the fallen leaves around the roses, revealing more snails, all of which I tossed into the yard waste container.

As you check your garden for snails, be sure to inspect your tomato plants for other pests. Aphids and tomato hornworms are prevalent this time of year. Wash off aphids or spray them with insecticidal soap. If it looks as if something has been eating your tomato plants, you'll probably find a fat green tomato hornworm, the same color as the foliage. These creatures are voracious. Put on a pair of gloves and hand-pick them off or lift them with your trowel, then discard them. If you've found one worm, look closely; there are probably others. Keep an eye on that tomato plant as you make your daily rounds.

Near the end of the month, my neighbor Rhonda brought over Sluggo, a new European snail bait that's supposed to be safe around pets and breaks down into fertilizer. Tired of hosting endless rounds of beer parties in the snail traps while other snails are devouring my plants, I dashed out to a store that stocked the new snail bait, bought myself a jug full, and sprinkled the bait throughout the yard.[5]

### Planting Herbs, Flowers, and Honeysuckle

JUNE IS A good time to plant herbs and summer annuals, and your last chance to plant most warm-weather vegetables: corn, beans, squash, melons, tomatoes, peppers, carrots, and beets. You can also plant tubers for dahlias, gladioluses, and summer-blooming bulbs. To fill the bare spots in the iris and herb gardens, I've sowed and set out more herbs and annuals. Earlier this month I sowed sweet alyssum seeds around the front gardens. In the herb garden I set out two six-packs of lobelia and some lemon thyme, purple sage, and basil. One advantage with herbs is that they're easy to grow in almost any soil. Another advantage is that many herbs don't appeal to snails and slugs. I've never had any problem with snails eating my rosemary, oregano, sage, and thyme.

Most of our favorite herbs have been around for centuries. Basil (*Ocimum basilicum*) was used in the Middle Ages to flavor sauces and drinks. It was also used as a strewing herb to sweeten the home well into the Renaissance. Sixteenth-century herbalists believed that basil helped cheer the heart.[6] It certainly adds cheer to many Italian recipes.

Sage (*Salvia officinalis*, from the Latin *salvare*, which means "to heal") was used in medicine and cooking by the Romans, who brought it with them to Britain. In the Middle Ages people used it, as we do today, to flavor poultry dishes and soups. Medieval and Renaissance men and women also used it for strewing, mixed it in potpourris, chewed it to clean their teeth, and made it into syrups to heal sore throats and lotions to soothe aches and pains. Believing that sage promoted longevity, they enjoyed sage ale and sage tea as health drinks.[7]

Thyme (*Thymus*) was believed to inspire courage. Medieval ladies often gave their knights scarves embroidered with thyme. It was also cooked as a pot herb in soups and pottages, steeped and drunk as a tea to aid digestion, used as a cold medicine, and strewn around the house. In the Renaissance thyme was made into a tonic to cure depression.[8]

Sometimes we choose to grow plants for symbolic reasons. This month after I turned in my final grades, marking the end of another academic year, I stopped at the local nursery. There I picked up a honeysuckle vine (*Lonicera japonica* Halliana) to replace the hibiscus that died last winter. The honeysuckle wasn't a very large plant, only a one-gallon size, for I had to fit it into my car, and the nursery's selection wasn't the best. The poor plant was pot-bound and looked rather the worse for wear. But I wanted a honeysuckle as a surprise gift for Bob, who associates its fragrance with fond memories of summers in Knoxville, Tennessee, where he went to graduate school. I watered the vine well and planted it in the bare space near Bob's study window. I've been carefully watering and feeding it, glad to see that the small plant has already put out some new leaves and a few clusters of flowers.

In the garden gifts return to the giver in unexpected ways. This Wednesday night we opened all our windows to cool the house after an especially hot day. As I sat reading in my study after dinner, I found the gentle breeze sweetened by the faint scent of honeysuckle wafting through the soft summer air.

## GARDEN REFLECTION

### *Enjoying the Present: Eating the Peach When It's Ripe*

OUR GARDENS OFTEN bring us important reminders. Last night a windstorm blew through the yard, knocking the biggest, most beautiful peach out of the dwarf peach tree. I had watched it ripen and was waiting for the perfect moment to pick it.

This morning I found it on the ground, smashed and eaten by snails. I had waited too long and missed the chance to enjoy it.

I realized that there was a lesson here for me. For most of my life I've worked hard, deferred gratification, postponing pleasure until I'd finished just one more project, one more piece of work. Deferred gratification can make us productive and successful, but it can also be carried to extremes, making us miss the pleasures right in front of us. The lesson I learned today is another version of presence: to pick the peach when it's ripe, not to defer gratification indefinitely.

I picked the remaining ripe peach and brought it into the house to enjoy tonight, sliced with frozen yogurt.

In many ways the warm days of June affirm for us the lesson of presence: gathering our rosebuds, recognizing opportunities, enjoying the present moment. Is there something you've been wanting to do but keep putting off, pushing it into the distant future: a trip to the beach, a short vacation, a phone call to a friend, or a special dinner with the one you love? If so, let the bright month of June, with its long hours of sunshine, light the way to a new realization and become a month of opportunity for you. Find a way to pick the peach and enjoy it when it's ripe.

# SEVEN

## July:
## The Fruits of Summer

Here's flowers for you:
Hot lavender, mints, savory, marjoram,
The marigold, that goes to bed wi'the sun,
And with him rises weeping: these are flowers
Of middle summer.

WILLIAM SHAKESPEARE,
*The Winter's Tale,*
IV.3, ll.103–7 (1611)

WHEREVER YOUR GARDEN grows, July brings all the bounty of summer. This morning I took a long look at the herb garden in my front yard. The colors are delicious. The chamomile blooms like tiny white daisies as the alyssum, lobelia, and nasturtiums weave their colorful patterns through a varied tapestry of green.

Gardening attunes us to the vast range of the color green. The single word seems too poor to comprehend all its different hues. The pineapple sage in the herb garden is olive green, stippled with yellow-green or chartreuse, while the nearby purple sage combines a darker green with shades of eggplant. The grass and clover along the cobblestone path are a bright spring green, while the trees surrounding the house are forest green, moss green, sea green, and blue-green. Backlit by the sun, their leaves are bright lime green, jade green, emerald green, luminous as stained-glass windows. Deep in the shadows, other greens recede into the colors of night. Medieval theologians saw green as the color of rebirth and eternal life, believing that it nourished both body and soul.[1] On this July day, surrounded by all these shades of green, I would heartily agree.

## GARDEN GROWTH

IN MY KITCHEN garden I've been harvesting salad greens and bok choy, and the tomatoes are finally ripening. The Early Girls are turning a rosy gold, and we had some sweet cherry tomatoes in our salads this week, the first of a long summer harvest.

The roses out front provide abundant color, fragrance, and fresh bouquets. Nearby, the first dahlias are blossoming in lilac and pearl pink. In our backyard the rose of Sharon (*Hibiscus syriacus*) by my study window is filled with violet blossoms, and the hydrangeas (*Hydrangea macrophylla*) bloom around the deck in shades of pink, lilac, and white.

Between my house and Rhonda's, the sweet peas (*Lathyrus odoratus*) are now over six feet high, covering the redwood fence with blossoms of scarlet, magenta, rose, and lilac, while summer bulbs provide an ongoing display of violet, orange, and gold. The purple-flowering spikes of liatris (*Liatris spicata*) are echoed by the blue-violet blossoms of lilies of the Nile (*Agapanthus orientalis*), Rhonda's violet society garlic (*Tulbaghia violacea*) and the soft lilac shades of an early-blooming dahlia. Fiery orange crocosmia (*Crocosmia masonorum*) blooms profusely throughout the garden, accented by dozens of golden California poppies (*Eschscholzia californica*). Like the fireworks displays

*First* April, *she with*
   *mellow showrs*
*Opens the way for early*
   *flowers;*
*Then after her comes*
   *smiling* May,
*In a more rich and sweet*
   *aray:*
*Next enters* June, *and*
   *brings us more*
*Jems, then those two, that*
   *went before:*
*Then (lastly)* July *comes,*
   *and she*
*More wealth brings in,*
   *then all those three.*
   ROBERT HERRICK,
   "The Succession of the
   foure sweet months"
   ll. 1–8 (1648)

this month, new bulbs provide daily surprises. Violet and peach gladio-luses (*Gladiolus grandiflora*) have begun to bloom, and a white summer amaryllis (*Crinum powellii*) came up last week, followed by exotic three-petaled tiger flowers (*Tigridia pavonia*). Lasting only a day, these tiger-striped red or yellow blossoms pop up in new spots all around the garden.

Bob and I have been eating ripe strawberries from our garden and blackberries from Rhonda's yard. We've sliced fresh lemons from our tree to flavor tall glasses of afternoon iced tea and add zest to main courses and salads. As you and I relax in our gardens and enjoy the fresh fruits of sum-mer this month, we participate in a long tradition. In July medieval and Renaissance men and women spent hours in their *herbers* or pleasure gar-dens, socializing and sharing outdoor meals, surrounded by fragrant herbs and flowers. In delightful contrast to their spare winter fare, in summer they had fresh fruit in abundance. From May until August people would enjoy apples, pears, plums, quinces, peaches, cherries, blackberries, gooseberries, raspberries, and strawberries in season. Oranges and lemons, brought to England during the Crusades and grown in palace gardens, gradually became part of their daily life. Medieval paintings and tapestries featured orange trees in blossom as symbols of fertility, and people used lemons to sweeten the breath, cure itchy skin, and lighten the complexion. Citrus juice was used in the Renaissance to cool the humors, fortify the heart, and reduce fevers. As more people began growing oranges and lemons, green-houses were developed in the seventeenth century to protect these delicate citrus trees from winter's chill.[2]

> The World *is a great* Library, *and the* Fruit-*trees are some of the* Bookes wherein we may read and see plainly the *Attributes of* God *his* Power, Wisdome, Goodnesse.
>
> RALPH AUSTEN, *A Treatise of Fruit Trees* (1653)

July's feast days celebrate gardens and the fruits of summer. July 11 was the Feast of St. Benedict, who established the first monastery gardens in sixth-century Italy. July 15 was St. Swithin's Day, celebrated in medieval England with elaborate meals, dancing, singing, and bobbing for apples. On July 25, St. James's Day, the gardener at Westminister Abbey would give ripe fruits from the garden to the community.[3] Equating the bounty of summer with God's beneficence, seventeenth-century devotional writers praised fruit trees as gifts from the divine creator.

Wherever you live July is a time to enjoy the ripeness of summer. The fresh fruits and vegetables, colorful flowers, and long summer days call us outside. July is a time to entertain friends and family with outdoor picnics and barbecues or spend a few quiet moments in your garden relaxing, sipping iced tea and listening to the birds. As you cultivate your garden this month, take time to enjoy the gifts of summer.

## GARDENING AS SPIRITUAL PRACTICE

### *Knowing When to Harvest*

THIS WEEK I harvested a bowl of ripe strawberries and Rhonda brought over some young zucchini from her garden. Harvesting fruits and vegetables teaches us a lesson in timing. If I pick my strawberries too soon, they're not sweet enough. If I wait too long, the birds will get them before I do or they'll rot on the vine. As for zucchini, if gardeners don't keep up with the harvest, a finger-sized zucchini will become the size of a baseball bat and just about as appetizing.

Timing is essential in the gardens of our lives. There are seasons and cycles in everything we do. Wisdom comes when we recognize these cycles. Otherwise, we can pick the fruit before it's ripe: act on impulse, finish projects too soon, or rush into a new relationship without getting to know the person first. Or we can wait too long to harvest: keep projects, clients, or friends "on the vine," waiting too long to take action or get back to them. We can also miss the rich time of harvest by procrastinating. As a teacher I know that when students feel inadequate they often put off studying or working on papers until it's too late to do a good job. Then their poor performance only reinforces their sense of inadequacy.

## Questions to Consider

KNOWING WHEN TO harvest is a valuable skill. Do you need to work on your timing? Ask yourself the following questions and see.

- Are you in such a hurry to finish things that you miss important details from the people and situations around you?
- Or do you put off harvesting until too late? Do you postpone working on challenging projects? Fall behind returning phone calls and e-mail? Are you out of touch with friends? A chronic procrastinator?

The keys to better timing are focus and mindfulness. Don't let yourself become distracted by the incessant noise and activity around you. Keep in mind what's important. At first you may need to set aside a day to catch up: to finish that project or return all the phone calls or e-mails. Then cultivate a system. Set aside time for what's really important, and do these tasks first. But also schedule regular time to return phone calls and keep up with correspondence so you don't get backlogged.

When I was a grad student at UCLA, one of my professors would go back to his office every day after his classes and research. Like clockwork he would be there at 4:00 P.M., tapping away at his typewriter. For the next hour he'd handle his correspondence, keeping up with friends and former students around the world. Then promptly at 5:00 he'd leave the office, go home to tend his garden and enjoy his life. He taught me an important lesson: if we set aside time to cultivate something on a regular basis, it gets done.

## Cultivating Greater Mindfulness

SOMETIMES, EVEN IN our gardens, we can move through our tasks mechanically, mindlessly. The other day I was trimming back the yellowing iris leaves into short fans when I suddenly realized that what I was cutting was not an iris but a gladiolus planted among them. An important lesson for me this year is getting acquainted with all the plants growing around me. When she planted the friendship garden with Rhonda, Betty Johnson mixed her spring and summer bulbs so that something is always blooming there. I stopped short, sparing the gladiolus, realizing how essential mindfulness is in a garden.

Mindfulness involves timing, for the right action at the wrong time is *not* right action. Cutting back the iris leaves when they've begun to turn brown and die makes sense. Cutting back a gladiolus at the same time makes no sense at all. I carefully staked the gladiolus stalks and stopped my trimming for the day.

Becoming more mindful means looking beyond rushing and routines enough to make sure our actions make sense in the larger scheme of things. I've now developed an increased respect for gladioluses. Two lovely peach and violet ones are blooming in my garden, reminding me to slow down and notice what's about to blossom in my life.

But what about the garden of your life?

- Is there something about to blossom now that you might miss if you don't take time for mindfulness?
- Could someone you know—a friend, family member, even yourself— benefit now from your care and attention?

This week take the time to be more mindful and see what you discover in your life's garden.

## What Nourishes You?

WHEN WALKING THROUGH your garden, you'll notice many things, including when some plants are suffering from chlorosis (iron deficiency). Recently the leaves of one of my potted azalea plants looked pale and washed out. Only the veins remained a healthy green. The poor plant was anemic: undernourished, weak, and run down. After I fed it a dose of chelated iron, its leaves turned a healthier green.

This month I've noticed that lots of my flowers and vegetables need more nourishment, so I've been fertilizing them. Moving from the garden around you to the garden within, July is a good time to consider your own need for nourishment: for appropriate rest, relaxation, and revitalization.

You probably find renewed energy and peace of mind in your garden, listening to the birds, enjoying the summer blossoms, walking among the herbs and flowers. But as your plants' needs differ according to their growth cycle, your needs, too, differ at different times.

*Rest: If you've been rushing around at a frantic pace, multitasking, trying to do too many things in too little time, you're probably exhausted.* You can nourish yourself now by simply resting.

- Put some lavender under your pillow and take a nap. Like too many people in our culture, you may just need more sleep.

*Relaxation: If you're more tense than tired, you need to relax and release your tension.*

- Do some stretching exercises.
- Take a slow, leisurely walk, becoming more mindful of your breathing, moving in time with your inner rhythms.

*Revitalization: If what you're feeling is not exhaustion but emotional fatigue, you need something to energize you.* Caregivers and people in the helping professions are especially prone to feeling drained. Take a break to revitalize yourself.

- Put on some of your favorite upbeat music.
- Get some fresh air. Go running or do some other aerobic exercise to get your circulation going.
- Check out some new cultural or educational activity in your area.

Last week I set myself a goal to finish a writing project by Thursday morning. Then I took the afternoon off to drive to *Sunset* magazine's Menlo

Park headquarters, where I walked around their display and test gardens, discovering new flowers, fruits, and vegetables that may appear in future magazines.[4] This welcome break revitalized me and lifted my spirits.

### Personal Exercise: Restoring Your Energy

WHENEVER YOUR ENERGY is low, ask yourself, What nourishment do I need now? Determine whether what you need is rest, relaxation, or revitalization. Then give yourself the proper nutrients.

### Cultivating Excellence, Not Perfection

WITH ALL THE garden tasks this time of year, sometimes it seems we can't keep up. Yesterday I stood outside my front gate wondering what to do next. All I could see were imperfections. The roses needed to be deadheaded, the grass and clover were growing over the cobblestone walk, the sweet alyssum was leggy and overgrown, and dandelions were invading the herb garden.

Feeling overwhelmed is nothing new for conscientious gardeners. Our time is limited, the power of nature infinite. In Milton's *Paradise Lost*, even the Garden of Eden made Eve feel inadequate. She complained to Adam that their garden was growing out of control:

> . . . the work under our labor grows,
> Luxurious by restraint; what we by day
> Lop overgrown, or prune, or prop, or bind,
> One night or two with wanton growth derides
> Tending to wild.[5]

We'll feel as frustrated as Eve if we strive for perfection in our gardens or our lives. The best we can hope for is excellence.

Although their dedication to work is legendary, the seventeenth-century Puritans recognized the dangers of perfectionism, which they called "excessive zeal." In 1622 the Puritan writer William Gouge criticized workaholics who were "*too intensive* upon their businesses, even the affaires of their lawfull callings (for in good things there may be excesse). Herein many Students, Preachers, Lawyers, Tradesmen, Farmers, Labourers, and others offend." Aware of our human need for balance, Gouge condemned perfectionism as unhealthy, explaining that "some are so eager on their businesse, that they thinke all the time mispent, which is spent in nourishing and cherishing their bodies; and thereupon wish, that their bodies needed no food, sleepe, or other means of refreshing. These thoughts and desires are foolish and sinfull."[6]

For conscientious people from the Puritans' time to our own, it's not easy to release demands for control and perfection. But perfection in this world is impossible. Our gardens are part of an ongoing process, not a perfect, static ideal. Frustrating and beautiful, filled with continual challenge, change, and surprise, our gardens are metaphors for life. They force us to realize that we are part of a pattern far greater than ourselves. We can do our part: cultivating the soil, planting the seed, watering, weeding, and tending with care, but life itself must germinate the seed and bring our dreams to fruition. In this process we participate and cooperate but can never control.

## Questions to Consider

HAVE YOU BECOME frustrated in some area of your life because you're looking for perfection? If so, ask yourself the following questions:

- What was happening?
- What did I do?
- What did I expect?
- What is my part in the process? What can I realistically do?
- What parts belong to others, to nature, and to life?
- How can I do my part to the best of my ability, then release the rest to the larger process?

I realize that my garden and my life will never be perfect. What I can do is aim for excellence: doing the best I can in the time I have.

## Garden Tasks

### Harvesting Fruits and Vegetables

FRUITS AND VEGETABLES grow and ripen so quickly during these warm summer days that it can be a challenge to keep up with them. But harvesting summer squash when they're young and tender not only keeps them from becoming oversized and tough but also helps the plants keep producing more fruit. The same goes for beans. If you keep picking them, the plants keep producing. So multiply your pleasures this month by harvesting your fruits and vegetables often.

### Planting and Planning

IN JULY THE summer heat makes it difficult for new plants to take hold, but if you're careful about watering you can still set out perennials and warm-weather annuals, as well as many herbs and vegetables (late corn, beans, and tomatoes in zones 8 through 11, cool-weather crops for fall in zones 4 through 8). This month I got four English lavender (*Lavandula angustifolia*) plants and set them out in the iris garden, where they'll provide height and fragrance between the irises and the bedding plants. I missed the lovely fragrance of the English lavender growing along the cement walk at our old house in San Jose. That plant seemed to thrive on neglect, flourishing in the summer heat reflecting off the cement. It was so vigorous I had to keep cutting it back or it would grow across the walk. Then I'd wrap the clippings in cheesecloth and put them in my drawers and closets, bringing inside the sweet scent of summer. I look forward to harvesting the new English lavender once it takes hold.

In medieval and Renaissance England, people used lavender to perfume gloves and linens, make potpourri, and fill sachets to put in closets and beneath pillows. They washed with lavender water, cooked with lavender-flavored sugar, and enjoyed lavender syrups, cordials, preserves, and tonics. Widely used medicinally, lavender was thought to cure headaches, cheer the heart, and comfort the spirits.[7] Centuries later this fragrant herb is still used in aromatherapy to relieve stress and promote relaxation.

In addition to the lavender, I got some Italian parsley to grow in a pot. Parsley (*Petroselinum crispum*) has been a culinary herb for centuries. During the Middle Ages and the Renaissance it was put into poultices to heal skin infections and taken internally as a diuretic and cure for indigestion.[8]

In my vegetable garden I made one last attempt to grow green beans—I love the flavor of fresh green beans and the fun of picking them right before dinner. This time I sowed the beans inside to keep the soil moist during germination. On July 1, I got one more packet of seeds—bush beans this time; the pole beans were all gone—and soaked them overnight. Then I put them in peat pots in my greenhouse box and set it on the kitchen counter.

In just a couple of days the bean sprouts began to emerge. When I took the top off the box to give them room to grow, the beans grew four inches in one day! I carried the box to my study and set it by the window. As I worked at my computer, the beans grew another two inches. Perhaps the summer heat contributed to this remarkable growth; it was 100 degrees that day. Watching these beans, I saw the old tale of Jack and the Beanstalk coming to life. When the temperature cooled down, I set the box outside to harden off the plants and a few days later planted them in the garden. I fertilized them with fish emulsion and surrounded them with a barrier of the new snail bait. Perhaps we can still have green beans this summer.

High summer is a time for planning as well as planting. The Holland bulb catalog came yesterday, a delight to read, recalling the procession of early spring color: crocuses, grape hyacinths, daffodils, tulips, and hyacinths. How much of gardening involves planning, selecting bulbs now for fall planting to usher in the glories of spring. The garden is

---

**Making Lavender Sachets**

*To make your own lavender sachets, cut lavender sprigs in the morning, before the sun dries their oils. Take them inside and spread them on a screen or hang them in indirect light. When the sprigs are dry, place them on small squares of cotton cloth and wrap them into bundles. Tie a ribbon at the top of each bundle. Or sew up three sides of the square like a small pillowcase, then turn the case inside out, fill with lavender, and sew up the remaining side. Place the sachets in your drawers or closets. They also make nice gifts.*

nature's evolving tapestry and an ongoing education. I ordered some new spring crocuses as well as summer-blooming liatris and allium, all of which should arrive in time to plant in October.

## Watering

UNLESS YOUR PART of the country gets lots of rain this time of year, it's important to water regularly. Throughout most of California irrigation is essential. In the San Francisco Bay area, summers are hot and dry. The grass on the East Bay hills turns golden brown from lack of rain. Without watering my garden, too, would soon turn brown.

When we moved here last fall, there was an automatic drip watering system in back but we still had to water the front yard by hand. Each day when I brought another carload of boxes into the house to unpack, I would run around the front yard, moving the sprinkler and the garden hose, worried about providing enough water to this beautiful new garden.

Gardeners have been watering their fields for most of recorded time, and medieval monasteries had remarkably sophisticated irrigation and plumbing systems. In 1165, during the reign of Henry II, the Benedictine monk Prior Wibert set up a system that piped water from a nearby hill to the monastery gardens at Christ Church, Canterbury. From the gardens the water was piped into the bathrooms; it finally flowed out as wastewater into a ditch. Until excavations in 1989 the three-inch metal pipe, buried fifteen inches below the surface, could still carry water down to the site of the old monastery gardens.[9]

Encouraged by this venerable tradition, I resolved to install a new irrigation system in the front yard, so I called a landscape gardener recommended by my friend Chris. One Saturday the gardener began digging, laying pipes, and setting up the new

> *Then when the fiery Suns*
> *too fiercely play,*
> *And shrivell'd Herbs on*
> *with'ring Stems decay,*
> *The wary Ploughman, on*
> *the Mountain's Brow,*
> *Undams his watry Stores,*
> *huge Torrents flow;*
> *And, rattling down the*
> *Rocks, large moisture*
> *yield,*
> *Temp'ring the thirsty*
> *Fever of the Field.*
> JOHN DRYDEN,
> *Virgil's Georgics,*
> I, ll. 157–62 (1697)

watering system. He installed sprinklers to water the trees and shrubs in the inner courtyard, a pipe leading to the bean plants on the north side, and a combination drip watering and sprinkler system for the front yard. Each rosebush now has its own drip outlet, to water its roots without getting its leaves wet. The ferns, evergreens, and small bedding plants are watered by a combination of drip and sprinkler systems. Everything is connected to a timer inside the garage, which turns the system on at five in the morning, when the water soaks into the soil without evaporating. We still need to connect it to the tomato plants in the south yard and the rosebushes on the strip between the sidewalk and the street. But even now the new watering system is magic, freeing me from running around each day with the hose and sprinkler. Now I have more time for other essential tasks, like weeding, fertilizing, trimming, and cultivating.

If you're not happy with the irrigation system in your garden, now is the time to do something about it. If you're handy with tools, you can set up your own system with equipment from the hardware store. A good drip system and a timer will help you save time, conserve water, and ensure that your garden is watered regularly, even when you're not around. While our friends Cory and Kevin were on vacation in England this summer, their tomatoes, beans, and cucumbers were watered daily by the drip system Kevin had set up. Three weeks later they came home to a healthy garden with lots of ripe tomatoes.

## Feeding Your Plants

SINCE OUR WATERING system was still incomplete, when we spent a few days in Mendocino this month, we asked our neighbors to water our tomatoes for us. When we got back, Ron said our tomatoes looked "weak."

Weak? I thought. They seemed to be growing well. I'd given them compost and regular doses of fish emulsion. But Ron's vigorous tomato vines were reaching over the fence, filled with ripening fruit, while ours were sitting there in their pots, obviously outclassed. It was time to get some tomato food and read up on fertilizers.

I got the tomatoes some new water-soluble fertilizer, especially good for container gardening. Then I walked around the front garden looking at the flowers, which also looked a bit weak. I've always gardened organically, using lots of compost, avoiding chemical fertilizers and pesticides, but this is a new garden with new soil conditions. I realized that since I've lived here the only flowers I've fertilized are the acid-loving azaleas, camellias, and rhododendrons, and the roses, which are growing beautifully. The bedding plants I've mulched repeatedly with compost, but for now this was not enough.

Unlike chemical fertilizers, which go to work immediately, the organic nutrients in compost aren't readily available to plants. Unless you've been adding compost to your garden for years, your soil, like mine, may lack essential nutrients. I realized I had put my plants on a deprivation diet, so I got some water-soluble plant food to tide them over. It may be my imagination, but the next day they already looked more vigorous.

## Understanding Fertilizers

THE THREE MAJOR nutrients required by plants are nitrogen (N), phosphorus (P), and potassium (K). The three numbers on fertilizer packages list their percentages in this order (20-10-10 means 20 percent nitrogen, 10 percent phosphorus, and 10 percent potassium).

Nitrogen supports photosynthesis and promotes the growth of healthy foliage. Plants need higher amounts of nitrogen to produce new foliage. This is especially important for lawns and leafy green vegetables. Phosphorus helps plants grow strong roots, flowers, and fruit. Potassium also promotes stronger roots and fruit while improving overall vitality. If you're growing spinach and lettuce, you'll want a fertilizer high in nitrogen, whereas if you're growing flowers or tomatoes, you'll want higher amounts of phosphorus and potassium. Give a high-nitrogen fertilizer to your tomato plants and you'll get lots of nice green leaves but not many tomatoes.

Most chemical fertilizers are complete, containing all three nutrients in amounts listed on the label, although some nitrogen-only fertilizers are sold for lawns, and combinations of phosphorus and potassium are sold as bulb food. Except for timed-release varieties, chemical fertilizers work more quickly than their organic counterparts. Liquid forms are available immediately. But many chemical fertilizers are derived from petroleum products, and they can burn sensitive plants. Runoff from chemical fertilizers pollutes our drinking water, flowing through storm drains into rivers and lakes or seeping through layers of earth into our groundwater.

If you garden organically, enriching your soil over the years, you may not need fertilizer. A soil test will tell you if you should have additional nutrients, and there are many organic forms available. You can add nitrogen with compost, manure, and fish emulsion; phosphorus with bonemeal, hoof and horn meal, and rock phosphate; and potassium with leaves and wood ash. To garden in harmony with the earth, you'll want to use fertilizers judiciously, building up your soil with compost and organic nutrients.

*Tips for*
*Deadheading Roses*

As you cut or deadhead
your roses, you can help
grow healthier plants by:

- cutting at an angle
  above a five-leaf node
- always cutting above
  an outward-facing node
  so the new shoot will
  grow out, leaving the
  middle of the bush
  open and less
  susceptible to disease.
- picking up the leaves
  and trimmings to
  prevent disease
- watching for signs of
  mildew, rust, or black
  spot and taking
  appropriate action

## Trimming and Pruning

THIS MONTH IT'S time to trim back flowering perennials, after they bloom and before they set buds for the coming season. Last month I cut back the heavenly bamboo (*Nandina domestica*) and the tea tree (*Leptospermum*), giving me a better view of the garden from my study window. When the beautiful violet rose of Sharon (*Hibiscus syriacus*) stops blooming, I'll trim it back as well.

It's important to continue deadheading the roses and other summer-flowering plants to keep them blooming. I've been shearing the spent blossoms of the alyssum and trimmed back the dried-out stalks of the California poppies out in front. This has not only made the garden look better immediately but also stimulated the plants to produce new blossoms. I also trimmed the chrysanthemums so they wouldn't get leggy.

The north side of my garden could easily become overrun. Last week I took my garden shears and cut back the vigorous hardenbergia vine (*Hardenbergia violacea*) that was reaching into the nearby liquidambar tree and across the north garden walk. I also pulled up some of the spearmint (*Mentha spicata*) that was blocking the path in front of the garden gate. All mints (*Mentha*) are remarkably vigorous plants that root quickly, spread by underground stems, and can easily become invasive. Many people try to control mint's growth by planting it in pots or boxes. But watch it closely, for mint is strong and clever. It will reach out, sneak under fences and sidewalks, and spring up where you least expect it.

Although it's bold and assertive, I like the mint growing in our north garden. It thrives on neglect, covers the earth with its cool, fragrant leaves, adds flavor to iced tea, and makes a pleasant after-dinner drink. It may even provide my dog with some relief, since it grows outside her kennel. Medieval gardeners grew mint for many reasons, using it to flavor meat dishes and to heal coughs, sores, bee stings, headaches, and digestive disorders. They would also strew it about the house for its pleasant scent and reputed power to repel fleas.[10] English settlers found the plant so useful that in the 1600s they brought it with them to the New World.

**Brewing Your Own Mint Tea**

*Brew your own homegrown mint tea by pouring boiling water over dried spearmint or peppermint leaves: 1 teaspoon of mint for each cup of water. Steep for 5 minutes, strain, and serve.*

## Watching for Pests and Disease

AS YOU WALK around your garden, gathering roses and herbs, harvesting vegetables, or deadheading spent blossoms, watch your plants for signs of pests or disease. The other day I noticed the leaves curling up on one of the rosebushes: a sign of mildew. So I sprayed the bush with a fungicide that contains sulfur, with its characteristic rotten egg smell. I also tracked down the probable causes, aiming one of the new sprinkler heads, which had been spraying the rosebush, back at the bedding plants.

I've been battling snails in my garden all summer. Last month, when I complained to our friends John and Catherine over dinner, they offered to take some of them. "It's too cold for snails in the mountains where we live," said Catherine in her French accent. "We could use them for escargot." I don't know if they were joking or not. But finally the new snail bait is working. I see less snail damage, fewer snails wandering around, and the plants are slowly recovering from the onslaught. Best of all, this bait can be used around domestic animals, and it breaks down into iron phosphate, fertilizing the soil as it protects the plants.[11]

*The summer's flower is to*
*the summer sweet*
*Though to itself it only*
*live and die,*
*But if that flower with*
*base infection meet*
*The basest weed outbraves*
*his dignity;*
*For sweetest things turn*
*sourest by their deeds:*
*Lilies that fester smell far*
*worse than weeds.*
WILLIAM
SHAKESPEARE,
sonnet 94,
ll. 9–14 (1609)

### Staking and Mulching

TYING UP AND staking plants is an ongoing task of summer. A few days ago high winds blew down two of my tomato plants that had outgrown their cages, so I tied their wandering branches to the fence behind them. I'm glad I staked the gladioluses last week; otherwise the wind would have broken off their long, thin stalks. I also set up wire cages around all the dahlias in the front garden. Thanks to the new snail bait, they are finally flourishing. Next to green beans, snails must find dahlia greens a special delicacy.

It was over 100 degrees yesterday, so the only garden maintenance I did was add mulch to protect the plants' roots from drying out. I bought organic compost and cedar mulch at the hardware store, and spread the compost around the herb, iris, and friendship gardens and the cedar mulch around the miniature roses next to the sidewalk.

## Tips for Mulching Your Garden

MULCHING YOUR GARDEN insulates the soil, keeping it cooler in the summer and warmer in the winter. It also helps control weeds, improve soil texture, encourage earthworms, prevent drought, and conserve water.

* **In the summer** a good layer of mulch (two to three inches deep) keeps the soil cooler and helps retain moisture, reducing your use of water. Mulch keeps down weeds by preventing their seeds from germinating. Slowly decomposing, it improves your soil's texture. Some organic mulch (leaves, grass clippings, compost) adds nutrients to the soil. By keeping your soil cooler, mulch attracts earthworms, which cultivate the soil closer to the surface. Mulch in your vegetable garden also protects strawberries, melons, and squash from rot and disease.
* **In the winter** a thicker layer of mulch (six to twelve inches deep) insulates the soil in colder climates, helping protect the roots of roses and other plants. By maintaining a more consistent soil temperature, it prevents damage to dormant perennials from alternating freezes and thaws. Winter mulch breaks down gradually and can be worked into the soil in the spring.
* **Organic mulches** come in many forms: shredded leaves, compost, wood chips, grass clippings, straw, shredded bark, buckwheat and cocoa bean hulls, sawdust, hay, seaweed, pine needles, stones, and gravel. These mulches have different qualities: pine needles are acidic (use these with acid-loving plants); wood chips and sawdust remove nitrogen from the soil (so you need to supplement with additional nitrogen); and rocks and gravel absorb the sun's heat during the day, then release it, keeping the ground warmer at night.
* **Inorganic mulches** include black plastic (which keeps the soil warm and prevents weeds, but many find it unattractive), old carpet, and newspaper (which gradually decomposes). Most gardeners cover inorganic mulch with a thin layer of soil, bark, or chips.
* **When *not* to mulch.** As helpful as mulch is during summer and winter, there are times and places you don't want to mulch:

- Don't mulch in early spring. Wait until the ground warms up.
- Don't mulch newly planted seeds or tiny seedlings or they'll suffocate.
- Don't mulch too close to plants' stems or they will rot.
- Don't mulch where you have problems with rodents, snails, or slugs. Mulch can create the perfect hiding place for pests.
- Don't mulch over the rhizomes of bearded irises or they will rot.[12]

This morning I walked around the garden, tidying up and visiting with my plants. In the past few days gusty winds have blown lots of leaves down from the trees and scattered them around the garden, so here I am raking leaves in July. As I was picking up magnolia leaves, I asked myself, Why keep throwing all these leaves into the yard waste to be carried away with the trash and then buy mulch? Many of the leaves are already dry and brittle, so I began crushing them and putting them down as another layer of organic mulch around the garden. That way nothing is wasted and the cycle continues, with old leaves falling to the ground, then returning to enrich the earth.

### An Unexpected Visitor

LAST NIGHT BOB noticed some unusual stirring outside his study window. There, on the deck, calmly eating our cat's food, was an opossum, a strange-looking animal about the size of a cat with a long tail like a rat's, feet like a monkey's, an unruly gray coat, and a large white head that tapered to a point, ending with a nose like a pig's. Intent on eating, the opossum seemed undisturbed as we leaned out the window to watch him, talking and laughing about his strange appearance. Finished with the food, the animal walked over to the cat's water dish, lapping up the water like a dog. Then it raised its head and looked straight at us with its beady black eyes before heading down the steps and into the night.

I was amazed by this close encounter with a wild creature in our own backyard. It was a dramatic reminder that we share this earth's garden with many life-forms. Today I walked through the garden with a new awareness of all the lives, seen and unseen, that share our neighborhood.

## GARDEN REFLECTION

### *The Power of Beauty*

LOOKING AT THE abundant flowers, fruits, and vegetables in my garden this month, I realize how much we are nurtured by the beauty of nature. Renaissance Neoplatonists described beauty as food for the soul. The Italian humanists Marsilio Ficino, Giovanni Pico della Mirandola, and Baldassare Castiglione saw beauty as the means by which humans apprehend the divine. This divine sense of beauty is affirmed by the glory of Renaissance art, from Botticelli's graceful *Primavera* to Michelangelo's heroic *David* and his masterpiece of creation in the Sistine Chapel.

Beauty is a consolation for life's struggles and storms, renewing our hope, refreshing our spirits. In some deep way, we all *need* beauty. Joining the Neoplatonists, I would add beauty to Abraham Maslow's basic needs, without which no human life can rise above subsistence. And more than all the clever inventions with which we humans "civilize" our world, it is nature's beauty that has always inspired artists, poets, and philosophers.

When my husband, Bob, was growing up in the Brooklyn housing projects, his mother would carefully budget their money, saving all year so she could take her sons out of the oppressive city summers to a small bungalow in the Catskill Mountains. There they would look out at the fresh green hills, listen to the wind and the songs of birds. They'd spend their days swimming in the lake, gathering wild blueberries, telling stories around the campfire, looking up at the stars. Running free in the woods under the vast summer sky, they breathed in more than the fresh mountain air. The natural beauty refreshed their spirits. When they returned to the city, they could see beyond the limited view from their fourth-floor apartment, realizing that the world was filled with possibilities. The first people in their family to attend college, both Bob and his brother became scientists, getting their Ph.D.'s in behavioral neuroscience. Today in their research they continue to uncover the wonders of nature, exploring new possibilities while discovering moments of beauty.

In this month of high summer, find a way to nurture your spirit with the beauty of nature.

# EIGHT

## August: First Harvest

Come, Sons of Summer, by whose toile,
We are the Lords of Wine and Oile;
By whose tough labours and rough hands,
We rip up first, then reap our lands.
Crown'd with the eares of corne, now come,
And, to the Pipe, sing Harvest home.

ROBERT HERRICK,
"The Hock-Cart,
or Harvest Home,"
ll. 1–6 (1648)

THE ILLUSTRATION FOR August in a thirteenth-century French Book of Hours shows a man and woman out in the fields, cutting the wheat and binding the sheaves. In the medieval church calendar, the first of August was the feast of Lammas (loaf-mass), the blessing of loaves from the first wheat harvest, followed by the Feast of St. Clare of Assisi on August 11 and the Assumption of the Blessed Virgin Mary on August 15. Lammas marked a shift of seasons: the end of summer and the beginning of the autumn harvest. This month medieval men and women were busy harvesting beans, peas, rye, barley,

oats, and filberts, which ripened around August 20, St. Philibert's Day.
From the Middle Ages through the Renaissance, August was a traditional
time for feasting, craft fairs, and harvest celebrations.[1]

*In Harvest time, harvest*
*folke, servants and all,*
*Should make all togither*
*good cheere in the hall.*
Thomas Tusser,
"August's Husbandrie,"
(1580)

Yesterday I picked our first green beans of the season and served them for dinner. Steamed until tender,
they were well worth waiting for. Not only are beans a
favorite in my summer garden but they were an essential source of protein during the Middle Ages. Since
the Benedictine Rule decreed a vegetarian diet, early
medieval monasteries grew beans in abundance for
their daily pottage, the thick soup of beans, peas, and
green vegetables eaten by the monks. Pottage was a
dietary staple for lay men and women as well since
meat was eaten primarily by aristocrats. The medieval
scholar Umberto Eco has even seen a connection
between the increased consumption of beans and peas, which provided
more protein for the general population, and the advancement of
medieval civilization around 1000 C.E.[2]

Conspicuously absent from the medieval diet, corn, potatoes, and
tomatoes were brought to Europe from the New World during the Renaissance. This month Bob and I have been enjoying vine-ripened tomatoes
from our garden. One of our favorite tastes of summer, tomatoes are the
most popular homegrown vegetable in the United States, not only delicious but incredibly healthful, rich in vitamin C and lycopene, an antiox-
idant that reduces the risk of cancer and slows down the aging process.
But for centuries Western Europeans and American colonists believed
tomatoes were poison and grew them only as ornamental plants.

Tomatoes (*Lycopersicon esculentum*) are actually a fruit, native to Central and South America. They were first cultivated in Peru, where tomato
vines still grow wild in the Andes, bearing golden fruit the size of currants
and cherries. The Aztecs discovered the wild tomato vine growing in their
cornfields, called the fruits *tomatl*, and enjoyed them in salsa and fried
dishes centuries before the Europeans came. In the early 1500s Hernando

Cortés brought tomato seeds back to Spain, where people were fascinated by these small yellow predecessors of our tomato. They called the fruit *manzana* (apple) but believed them to be poisonous, growing them on trellises as ornamental vines.

By 1554 tomatoes had been introduced to Italy, where they were called *mala insana* (unhealthy apple) until some adventurous Italians began frying them in olive oil with salt and pepper, broiling them with spices, and using them in sauces. Rechristened *pomodoro* after the golden apples of antiquity, tomatoes soon became prominent in Italian cuisine.

Tomatoes then made their way to France where they became known as *pommes d'amour*, aphrodisiacs. In Renaissance France couples gave each other tomatoes as love tokens. But long after these fruits had become a staple in Mediterranean cuisine, the English continued to shun them, believing they caused illness and insanity. Tomatoes were not readily available in England until the 1880s.

Most American colonists believed the English rumors and avoided tomatoes. One exception was Thomas Jefferson, who had enjoyed *pommes d'amour* while serving as ambassador to Paris. He raised these French delicacies at Monticello as early as 1782, sharing them with his Virginia neighbors. Yet most Americans avoided eating the fruit until the nineteenth century. Even in 1802 an Italian painter tried in vain to persuade his neighbors in Salem, Massachusetts, to taste a tomato. Finally, in 1820, Colonel Robert Giddon Johnson publicly ate tomatoes on the courthouse steps in Salem, New Jersey, and survived, dramatically disproving the rumors.

Slowly tomatoes became acceptable, recognized as a food after 1860 and mentioned in several cookbooks of the Civil War period. In 1865 they even entered the records of the Supreme Court when it declared that, although technically a fruit, the tomato could be sold as a vegetable. By the twentieth century Americans were eating an average of thirty-six pounds of tomatoes annually. Vine-ripened tomatoes not only taste better than commercial varieties but they contain nearly twice as much vitamin C. So for reasons of health as well as taste, this month treat yourself to fresh tomatoes. If you haven't grown any yourself, visit a local farmers' market to enjoy the ripe taste of summer.

## GARDEN GROWTH

IN ADDITION TO producing green beans and tomatoes, August has brought new growth to other areas of my garden. The dahlias of late summer are now in bloom. One early bush in the friendship garden has been filled for over a month with a profusion of lilac blossoms. Joining the roses in front for a late summer display, other dahlia bushes are rising to the occasion, blossoming in lilac, pearl pink, and bright tangerine.

Although the lilies of the Nile, gladioluses, and liatris have finished blooming and the orange crocosmia blossoms are nearly gone, exotic red and gold tiger flowers are still popping up around the friendship garden. The violet blossoms of Rhonda's society garlic are still going strong, and golden California poppies blossom at intervals throughout the yard.

The front of the iris garden needs lots of water this month to keep the nasturtiums happy. The lobelia, chamomile, and white alyssum bloom on, joined by the violet alyssum I planted there in June. Along the north yard pink and white hydrangeas are still in bloom. In the back the honeysuckle has begun to grow up its trellis, and the rose of Sharon is filled with violet blossoms.

In the inner courtyard clusters of rose-colored blossoms have appeared on a small tree growing next to the St. John's wort. I looked it up in my *Western Garden Book* and learned it is a crape myrtle (*Lager-*

*stroemia indica*).[3] Native to China, this plant does well in many climatic zones. Between July and September its branches are filled with blossoms of red, rose, pink, orchid, purple, or white.

As we approach summer's end, this season's abundant growth can lead to new discoveries. Walking over to Vasona Lake this month, I saw a tomato plant growing in the back of an old Chevy pickup truck parked in the driveway of a vacant house. The house looks deserted. Its windows are boarded up; weeds and trash fill the front yard. Loaded with tires and debris, the old red truck has been sitting in the driveway as long as I can remember, its paint oxidized to dull orange, its license expired. Yet here was a tomato plant, rising from the tires lying in the bed of the truck. Apparently, the fallen leaves from last year had turned to compost inside a tire, creating an accidental planter. Somehow, a tomato seed had found its way there, germinated, and taken root, receiving just enough moisture to survive. If a tomato plant can affirm new life where we least expect it, what more is possible when we take time to cultivate the gardens within and around us?

## GARDENING AS SPIRITUAL PRACTICE

### Seeds of Late Summer

IN OUR GARDENS August is a time of endings and beginnings. This month, as we harvest our tomatoes, beans, and squash, many of us are planting cool-weather crops: lettuce, spinach, and beets. But wherever your garden grows, late summer is a good time to plant seeds of renewal and cultivate healthy new habits in your life.

This month I began a regular practice of taking a daily walk with my dog around the neighborhood and up to Vasona Lake. Walking is a simple exercise that can be done almost anywhere or any time without expensive equipment. Its benefits range from lower blood pressure, lower cholesterol, and stronger muscles and bones to better overall health both physically and mentally. A gentle, weight-bearing exercise, walking helps prevent osteoporosis as well as many back problems. It helps us overcome stress, sleep better, feel more energetic, firm up, lose weight, look and feel more youthful. Walking also reduces the pain and severity of arthritis and boosts the immune system, reducing the risk of heart disease, strokes, cancer, and diabetes. It even increases endorphin levels in our brains, making us happier by improving our outlook.

*Personal Exercise:*
*Beginning a Walking Practice*

JUST TWENTY MINUTES of brisk walking every day can make a major difference in your life. If you'd like to begin your own walking practice but can't find the time for regular daily walks, be creative.

- Park your car farther away and walk across the parking lot when you go to work or run errands.
- Walk to work or walk *at* work, taking a twenty-minute walk on your lunch hour. The exercise will help you clear your head and cope with stress yet still leave time for a healthy lunch.
- If the weather is bad, bundle up in layers or do what a friend of mine does: go walking in malls. But remember that your goal is walking at a steady pace, *not* stopping to shop.

*Personal Exercise:*
*Planting Seeds of Health*

STUDIES HAVE SHOWN that it takes about thirty days to establish a new habit. This month, while the summer sun still lights the days, choose one healthy habit you'd like to cultivate. It could be walking or another exercise practice, eating a healthier breakfast, drinking more water, eating more fruits and vegetables, or something else. Remember, in your life as in your garden, growth is incremental.

- Start small. These are seeds you're planting, not full-grown plants.
- Set a goal and then take your first step to get there. That way you'll establish momentum to keep you moving forward. If you begin a

walking program, start with ten-minute walks three times a week, then work up to longer walks and add more days.[4]

🐑 Whatever new habit you choose, congratulate yourself for planting the seeds for a healthier tomorrow.

## Cultivating Awareness

THIS MONTH THE crape myrtle in my garden taught me an important lesson: when we begin paying attention to something, we see a lot more of it. Whatever we give our attention to increases. After discovering this plant in my garden, I've started noticing crape myrtles all around me. Driving around town, I saw pink, rose, violet, and red crape myrtles blossoming up and down the streets. Today when I drove out to campus I saw a row of pink crape myrtles next to the parking lot. When I walked over to the library, two red crape myrtles greeted me by the front door. My world is suddenly filled with crape myrtles. But those trees were *always* there. The only difference is my awareness.

## Personal Exercise:
## Focusing Your Attention

REMEMBER: WHATEVER WE GIVE our attention to increases. Drawing upon this principle, you can make a positive difference in your life this month by simply paying attention to something you'd like to see more of.

🐑 First ask yourself: What quality would I like more of in my life: Joy? Vitality? Prosperity? Beauty? Some other quality?

🐑 Then begin *noticing* this quality. If you picked beauty, *look* for it. As you go about your day, slow down, really open your eyes and look around you. Find something beautiful. This shouldn't be too difficult. Beauty has many forms: a new flower in your garden, sunlight on the treetops, a sunset, a loving gesture.

🐑 When you've found the quality you're looking for, acknowledge it. Name it to yourself ("How beautiful!"). Take a moment to appreciate and be grateful. Then move on.

🐾 Repeat this practice, consciously looking for your chosen quality every day for a week. Make it a game you play with yourself.

🐾 Once you've begun the process consciously, you will become more attuned to the quality, finding more of it in unexpected places.

### Personal Exercise: Turning Your Attention from Negative to Positive

ANOTHER WAY TO perform this exercise is to pick something you *don't* like (for many people this is easier than thinking of something they do). Remember that what we give our attention to increases, so don't focus on the negative quality. If you can do something to improve the situation, take action. (For example, if you've been worrying about your car, take it in for service). But don't just keep worrying.

In this exercise you can use the quality you *don't* like as a valuable resource, transforming it into something you *do* like.

🐾 Now that you've found something you don't like, think of the exact opposite. For example, if you've been complaining about your job, think about what you dislike most and reverse your impression of it. Boring work becomes "enjoyable work"; a stressful, negative environment becomes "an energizing, upbeat environment"; an unpleasant boss or co-workers become "supportive colleagues." At first this exercise might feel ridiculous, but try it anyway. Pick one good quality that's the mirror opposite of what you dislike.

🐾 The next step is to *look* for the good quality. Your brain is not used to this kind of thinking and will initially resist. But make an effort. Search for the quality you want. If you'd like more likable colleagues, look for *something* in your colleagues to like. Keep trying. Make this a game for yourself.

🐾 When you've found something to like, name it to yourself ("I like the way Paul smiles" or "I like Marie's expertise with computers"). Smile in acknowledgment and move on.

- Look for the next experience to like and repeat the procedure. Do this consciously every day for a week.
- Make it a game to find at least one example of what you're looking for each day. Then keep score, carrying an index card with you and making a mark beside each day's date when you find your quality.
- Once you've begun, you will find yourself noticing the quality you like more often. Remember to name it to yourself and record it.
- Don't be surprised if you find more possibilities and opportunities opening up for you, moving you from the old situation into one filled with the quality you desire.

Focusing your attention is an important way of cultivating your inner garden. Centuries of spiritual traditions have taught that we can change the world around us by first changing the world within us. Try this exercise and see what new opportunities blossom in your life this month.

### The Great Mandala

WE'VE BEEN GARDENING together in the context of the Middle Ages and the Renaissance, but nature's lessons transcend times and cultures. As long as men and women have cultivated the earth, they've recognized the wheel of time, the cycle of the seasons. In England medieval and Renaissance Christians followed the seasons of the liturgical year. The Aztecs in Mexico had their elaborate solar calendars, the Native Americans their sacred hoop. The Chinese plotted the cycles of yin and yang, Africans lived by the natural cycles described in *The Song of Lawino*, and the Hindus and Tibetan Buddhists had their circle of life: the great mandala.[5]

In the gardens of our own experience, each of us rediscovers this circle of life. Last night after dinner with our friends Cory and Kevin, Bob and I drove to the municipal rose garden for an evening walk under the stars. We began walking there years ago, when we started going out. With over five thousand plants and two hundred varieties of roses, the rose garden is the heart of an old San Jose neighborhood with stately homes and mani-cured yards. It's the site of many celebrations—picnics, weddings, and graduations—and it's always been a special place for Bob. He prepared for his marathons by running daily laps around its half-mile perimeter. There he met Terry and Sandy, who became his running partners, training together for their first marathon in 1987.

I sing of Brooks, of Blossomes, Birds and Bowers:
Of April, May, of June, and July-Flowers.
I sing of May-poles, Hock-carts, Wassails, Wakes,
Of Bride-grooms, Brides, and of their Bridall-cakes.

ROBERT HERRICK, "The Argument of His Book," ll. 1–4, from Hesperides (1648)

As Bob and I walked around the perimeter on this summer evening, he pointed to some of the houses, telling me about the people who lived there, people he'd met on his daily runs and cooldown walks among the roses. There were friends, neighbors, fellow run-ners, familiar faces who'd become a community, culti-vated by daily contact as Bob ran up and down the four streets that enclose the rose garden.

That evening at dinner we'd discussed our col-league Ed's latest book of poetry, Works and Days. It's a beautiful book about cultivating the olive groves at his home in Italy. The poems' titles span the alphabet in Italian from ago (needle) to zappa (hoe), tools used in his daily labors, instruments with which gardeners fill their time and measure their lives in a recurrent cycle of planting, cultivating, and harvest. The book itself is a cycle, with the last poem ending where the first began: in a bilingual pun on ago, the needle that threads through the fabric of time.[6] Like Ed, centuries of poets have found inspiration in the cycles of nature. The Renaissance poet Robert Herrick wrote of seasonal feasts, celebrating the harvests of fruits, flowers, and wisdom within the circle of life.

In a culture structured by business and commerce, we perceive time as linear, a collection of discrete parts. Each year has a distinct number—1999, 2000, 2001, 2002—a clear beginning and end. But in the garden year each season connects to the next, circling back like the poems in Ed's book or Bob's laps around the rose garden. As gardeners we participate in the circle of life as we cultivate the soil again and again, adding compost, weeding, watering. Cultivating the soil is not linear, not something we can do once and be finished. Good soil is created layer upon layer, season upon season, with patient, repeated applications over time of compost and other organic matter. Sometimes our individual actions may seem unremarkable, but the cumulative effect is powerful.

So, too, do we cultivate relationships, friendships, and deep, abiding love: over time, by recurrent cycles, one small action connecting to the next. Bob and I paused at the end of our evening's walk to look up at the pine trees, towering over the houses, then to the constellations of stars above our heads, realizing that our lives and all we survey are part of nature's great mandala: the circles of life, above, within, and around us.

> *Thy firmness draws my circle just,*
> *And makes me end where I begun.*
> JOHN DONNE,
> "A Valediction Forbidding Mourning,"
> ll. 35–36
> (c. 1600)

## Questions to Consider

AS WE MAKE our way through life, wisdom is recognizing the cycles. Living reactively, people dash from one action to the next, but living creatively, we see how these actions fit together into larger patterns of meaning. As you consider the circles of life in your world, ask yourself these questions:

- What are some important cycles in my life?
- What do they cultivate: Love? Self-respect? Friendship? Community? More than one of these?
- What small actions combine to reinforce these cycles, cultivating greater depth and meaning?
- Is there something I can do to cultivate an important cycle in my life today?

## GARDEN TASKS

### *Planting and Planning*

THE GARDEN TASKS for August depend on the length of your growing season. There's still time to sow cool-weather vegetables in zones 5 through 11. However, in zones 5 and 6, you'll need a cold frame because the nights are already growing colder. In zones 9 through 11 you may need to start your crops indoors to protect tender seedlings from the hot August sun. This month I sowed beets in a raised bed on the south side of the house. The tiny seedlings came up almost immediately: amazingly beet red with tiny green leaves. Beet greens are great in salads, so thinning them should prove an enjoyable task. I also sowed more mesclun greens and spinach in planters in our front courtyard. These seedlings, too, came up in just a few days. In late summer it's fun to witness all these new beginnings. For my flower bed in the front yard, I sowed pansy seeds in my greenhouse box inside. The days are still so hot and the tiny seedlings so fragile I'm glad I didn't sow them in the soil outside. As they've gotten larger I've been taking them out for a few hours to harden them off. In a week or so, on a cooler day, I'll plant them in the iris garden.

### *Watering, Mulching, and Sunscreens*

IN SOME PARTS of the country, late summer can be quite hot and dry. This month I fed the nasturtiums in front with some chelated iron. The plants looked stunted, and their leaves were yellow, signs of chlorosis. But as I examined the garden more closely, I saw that these plants had another deficiency: water. I stuck my finger into the soil and it was bone dry. In our gardens nothing can be taken for granted, even automatic watering systems. Since my system goes on at 5:00 A.M., I haven't seen it operating lately, so that day I turned it on manually and walked around the yard, checking the spouts to make sure they weren't clogged, redirecting one nozzle away from a rosebush, reattaching a tube that had become disconnected.

If you have an automatic watering system, it's a good idea to check it periodically. You might also need to reset the timer to cope with changing weather conditions. Since it's been quite hot here this month, I doubled the setting of ten minutes to twenty. I'll inspect the garden again in a few days to make sure the plants are getting enough water. Later in the season, when we get our fall and winter rains, I'll set the system back and eventually turn it off.

About the middle of the month, I covered the front gardens with more organic compost. My repeated applications of compost this summer are doubly beneficial: they help insulate the soil from the hot sun now and will improve its texture and supply more nutrients later.

As the sun shifts with the seasons, some areas of your garden will get more shade while others will be warmer. I've been mulching the green beans in the north garden and giving them additional water in the afternoons. The sun has been so intense that even with the twenty-minute daily soakings, their soil was parched by late afternoon. The other day the poor plants looked visibly stressed, holding their leaves up vertically to lessen the effect of the sun's burning rays.

Sun protection is important for all of us. When I go out into the garden, I wear a hat and sunscreen. Realizing that the beans might benefit from similar protection, I rigged up a portable sunscreen, placing a wooden clothes drying rack next to the plants and spreading a couple of old towels over it to filter the sun in the afternoons. The next day the beans looked much less stressed.

*Now carrie out compas,*
*when harvest is donne,*
*where barlie thou sowest,*
*my champion sonne:*
*Or laie it on heape, in*
*the field as ye may,*
*till carriage be faire, to*
*have it away.*

*Whose compas is rotten*
*and carried in time,*
*and spread as it should*
*be, thrifts ladder may*
*climbe.*

THOMAS TUSSER,
"Worke after Harvest"
(1580)

## Feeding Your Plants

IN LATE SUMMER it's important to feed the plants still blooming in your garden, those forming buds and bulbs for later blooms, and the vegetables you're growing now. This month I felt like a short-order cook. I continued to feed the tomatoes regularly as well as the annuals and dahlias. I put out some bulb food for the irises, citrus food for the lemon trees, all-purpose fertilizer for the honeysuckle and peach tree, and acid food for the rhododendrons, azaleas, and camellias, which are already forming buds for next spring. I gave the first of three feedings to the chrysanthemums to promote lots of fall blooms and put the last set of fertilizer stakes out for the roses. The rule is to feed the roses one last time six weeks before the first frost, which, of course, depends upon your climatic zone. In zones 5 through 11, roses will most likely bloom on for another month while in zones 1 through 4 it is already time to trim them back and cover their roots with extra soil for winter protection.

## Harvesting, Trimming, and Pruning

THIS MONTH IT'S important to keep up with your harvest, picking fruits and vegetables as soon as they're ripe, as well as deadheading the plants that are still blooming. I've been harvesting beans, fresh greens, strawberries, and tomatoes, and deadheading the roses and dahlias. I've trimmed back the spent blossoms of some hydrangeas, and have been cutting back the California poppies after they've gone to seed. The result was renewed vigor, more fruit, and blooms. As the iris leaves gradually die back, I've been trimming them into fan shapes and removing the dried foliage. All of the trimmings and deadheaded flowers go into the compost bin.

If some of your early spring vegetables have died back, it's time to clean up their debris and add that to the compost bin as well. If your strawberry plants have stopped bearing, this is a good time to trim off any unwanted runners and dead leaves, take out any diseased plants (discard and do not compost these), remove the old straw, and enrich the soil, cultivating it and adding compost. Since our plants are still bearing fruit, I'll wait a few more weeks to do this.

## Watching for Pests and Disease

EXCEPT FOR OCCASIONAL needs to spray fungicide on the roses to catch a bit of mildew on their leaves, this month my garden has been blessedly free from pests and disease. May you experience similar good fortune in yours. If not, the best policy is to take action as soon as you notice anything amiss. Control slugs and snails with bait and traps, and use insecticidal soap and fungicide as needed for harmful insects and mildew.

## An Herbal Lesson

THIS MONTH I'VE been harvesting herbs for seasoning and salad: cooking with fresh thyme, basil, and rosemary, and sprinkling dill on new potatoes and sliced garden tomatoes. This year's dill was a proud performer in my garden. By late July the plants' high, thin stalks were taller than I am and filled with lots of aromatic foliage. In the Middle Ages dill was a favorite, used to flavor fish and pottage, its name derived from the Anglo-Saxon word *dilla*, which means "to lull." Medieval herbalists believed that dill calmed the spirits and its seeds aided digestion, and dill is still a favorite herb in Scandinavian cuisine.[7]

Many of my summer menus have been inspired by the vigorous dill plants. Each evening I would take my kitchen shears out to the south garden to snip some fragrant sprigs for summer salads and suppers. But this week I was stunned to find the mighty dill knocked down to half its height, humbled by our afternoon winds. I carefully inspected the bent stalks. They were bruised but still intact, so I gently set them upright and staked them, borrowing a steel spiral from the tomato vines and carefully taping them in place. Then I watered them well and hoped for the best.

In our gardens there are always unexpected occurrences. The tallest herb in the garden can be knocked down by one gust of wind. Seventeenth-century poets would see this episode as a lesson in humility. With shorter plants the wind presents few problems. But aspiring to greater heights can be risky without support.

I watched the dill carefully for the next few days, and it seems to have made it through its accident. It is less majestic than before, staked and secured to the steel spiral with green garden tape, but it reminds me that when we reach for new heights, independence is an illusion. Even the brave need faith and support.

## GARDEN REFLECTION

### Friendship, Feasts, and Harvest Cycles

IN THE MIDDLE Ages people gathered in August to celebrate the first harvest. With all the fresh fruits and vegetables this month and the days still blessed with summer sunshine, August is a good time to share a festive meal with friends.

Elizabeth and Bill are old friends and mentors from my early days at Santa Clara. A Jesuit priest, Bill was once president of the university. Elizabeth was the first woman tenured in my college, the first woman president of the faculty senate, and, later, the first director of faculty development. They have both moved on from Santa Clara. Retired from teaching, Elizabeth now works with the American Association of University Women to help the homeless. After serving as president of two universities, Bill is now director of the beautiful Jesuit retreat house El Retiro San Iñigo in Los Altos. Bob and I hadn't seen either of them for quite a while, so we invited them over for dinner.

They arrived at our front door with smiles and a special housewarming gift from Elizabeth: a lovely bonsai tree, which occupied the place of honor as our centerpiece that night. The next day, while reflecting on our enjoyable evening, I studied the new bonsai, wondering just what kind of tree it was.

Everything about the tree is a perfect miniature. Standing ten inches high, in a seven-inch jade green pot, it looks like a full-sized tree, only proportionately smaller. Its trunk curves into a graceful spiral, as if it has danced with the wind. The small leaves are in perfect proportion to its trunk and branches. At its base the ground is covered with delicate moss

These are thy wonders,
  Lord of love.
To make us see we are but
  flowers that glide:
Which when we once can
  finde and prove,
Thou has a garden for us,
  where to bide.
Who would be more,
Swelling through store,
Forfeit their Paradise by
  their pride.
  GEORGE HERBERT,
    "The Flower,"
      ll. 43–49
        (1633)

and a tiny gravel stream. Beneath the tree, near a rocky ridge, representing distant mountains, stands a tiny statue of a wise old Buddhist priest with a scroll in his right hand.

Later that week I visited the nursery in Los Gatos where Elizabeth had gotten the tree. The bonsai gardener remembered Elizabeth and knew the plant well. "It's an elm," he said, explaining what kind of air, light, watering, and food it preferred. "Water it every day. Fertilize it with Miracle-Gro twice a month. You can leave it inside by a sunny window or outside this time of year in filtered sun. When it gets too cold, bring it inside. Oh, yes," he added with a smile. "It's deciduous. It will lose its leaves this fall."

I thanked him for the information and put the small tree in our courtyard beside the front door. There it gets lots of fresh air and is sheltered from the sun by the Japanese maple. Every morning when I go out the front door, I give the tree a drink of water. It seems happy in its new home.

Bonsai trees like this are at least twenty years old, paralleling my friendship with Elizabeth, which goes back over twenty-five years. For over two decades the tree has lost its leaves in the fall, remained dormant each winter, and burst into new life in the spring. Like a good friendship, it has known many cycles of challenge, change, loss, and renewal, growing stronger and more beautiful with them all.

> *Love, all alike, no season*
> *knows, nor clime,*
> *Nor hours, days, months,*
> *which are the rags of*
> *time.*
>
> JOHN DONNE,
> "The Sunne Rising"
> ll. 9–10
> (c. 1600)

Strong and beautiful as bonsai trees, friendships give shape and meaning to our lives, helping us mark the time in ways unknown to many in a world where digital watches measure time in one-second flashes. Yet beneath all the layers of technology, the sense of the sacred still remains in moments of love and inspiration.

Friendship brings our lives greater depth and meaning. We discover more of ourselves through our friends' eyes, as we share our ideals, joys, and frustrations. Friendship helps us make sense of an often confusing

and chaotic world, taking us into a larger reality that embraces many summers, autumns, winters, and springs.

   This month, as we begin the harvest season, remember to reach out to
the people you care about. Get in touch with an old friend: send a card,
make a call, share a meal. Take time to renew the enduring gift of friendship.

# AUTUMN

# Autumn Garden Checklist

## EARLY AUTUMN

🌿 **Plant:** Plant evergreens, trees, shrubs, perennials, and spring bulbs. If your winters are mild, store bulbs in the refrigerator for six weeks before planting; sow and set out cool-weather annuals and vegetables. If your winters are cold, transplant tender geraniums, begonias, and herbs into pots to overwinter inside.

🌿 **Water:** Water regularly if it does not rain often in your region, giving special attention to container-grown plants and any new plantings. If fall weather is warm, give tree roots deep soakings.

🌿 **Weed:** Clean up weeds and plant debris. Weed vegetable gardens well at the end of the growing season.

🌿 **Feed:** Feed cool-weather vegetables. Fertilize chrysanthemums one last time before their buds open. Add bonemeal when you plant spring bulbs. Feed roses up to six weeks before the first frost.

🌿 **Cut back:** If your strawberries have stopped bearing, trim off old leaves and unwanted runners. Take out unhealthy plants, remove the old straw, and cultivate the soil between rows. Clean up your garden, composting spent annuals and vegetable plants. Deadhead roses and other flowers still in bloom. Prune climbing and rambling roses when they've finished blooming. Wait until winter to prune trees and shrubs. In colder areas, dig and store dahlias, gladioluses, and other tender bulbs inside over the winter.

🌿 **Cultivate:** Rake and compost fallen leaves. Don't let them suffocate ground cover and bedding plants. As nights grow colder, cover tender vegetables. Dig and cultivate empty garden areas. Clean and store garden stakes. Divide and replant overcrowded perennials when they've finished blooming. Continue dividing and replanting irises, completing the task four to six weeks before the first frost. If your winters are cold, trim back roses and mound soil around their bases to protect the roots. Cover each mound with deep mulch after the first frost. Mulch perennials and around the bases of trees and shrubs to prevent frost damage. Stake young trees to provide support during winter storms. Root cuttings of herbs, geraniums, and other annuals to overwinter indoors.

✂ ***Harvest:*** Harvest the last corn, tomatoes, and other summer vegetables. If frost threatens, bring in green tomatoes to ripen inside. Harvest apples, pears, and pumpkins. Dry and store onions, potatoes, and other root crops. Dry herbs to put away for winter.

✂ ***Watch for:*** Pick up fallen leaves, blossoms, and fruit to discourage pests and diseases. Discard diseased fruit—do not compost it. Watch for pests and diseases and treat as needed.

✂ ***Enjoy:*** Enjoy the last roses of summer and the golden days of early autumn.

## MID-AUTUMN

✂ ***Plant:*** Before the ground freezes plant bare-root roses, trees, bushes, and hedges. Plant spring bulbs. If your winters are mild, store bulbs in the refrigerator for six weeks before planting; plant cool-weather vegetables (check seed packets for the correct planting time for your area). If your winters are cold, transplant tender geraniums, begonias, and herbs into pots to overwinter inside.

✂ ***Water:*** Water regularly if it does not rain often in your region, giving special attention to newly planted and container-grown plants. Deep-water young trees and new plants before the ground freezes.

✂ ***Weed:*** Clean up weeds and plant debris.

✂ ***Feed:*** Feed cool-weather vegetables. Add bonemeal when you plant spring bulbs. Feed roses until six weeks before the first frost. As their growth cycle slows down, stop fertilizing trees and shrubs (new growth could be vulnerable to frost damage).

✂ ***Cut back:*** In colder areas dig and store dahlias and other tender bulbs inside over the winter. Keep deadheading roses and other flowers still in bloom. Cut back perennials when their tops die down. Prune climbing and rambling roses when they've finished blooming. Clean up your strawberry patch, eliminating old leaves, unwanted runners, and unhealthy plants. Remove the straw and cultivate the soil between rows. Remove and compost spent annuals and vegetable plants. Wait until winter to prune trees and shrubs.

✂ ***Cultivate:*** Rake up and compost fallen leaves. As nights grow colder, cover tender vegetables. Clean and store garden stakes. Dig and cultivate empty garden areas. Divide and replant overcrowded perennials. If your winters are cold, trim back roses and mound soil around their bases to protect the roots. Cover each mound with deep mulch after the first frost. Mulch around perennials, trees, and shrubs to prevent frost damage. Stake young trees to provide support during winter storms.

✄ *Harvest:* Harvest apples, pears, raspberries, pumpkins, and other fall fruits and vegetables. Dry and store onions, winter squash, potatoes, and other root crops. Harvest late vegetables, and enjoy any tomatoes that have ripened inside. Dry herbs to put away for winter.

✄ *Watch for:* Check for slugs and snails. Treat with traps or bait. Watch for mildew or fungus and treat with fungicide.

✄ *Enjoy:* Enjoy the rich harvest, colorful foliage, and chrysanthemums of autumn.

## LATE AUTUMN

✄ *Plant:* Before the ground freezes, plant spring bulbs and bare-root roses along with bare-root trees, bushes, and hedges. Plant deciduous trees after they've lost their leaves. If your winters are cold, transplant tender geraniums, begonias, and herbs into pots to overwinter inside. If your winters are mild, set out cool-weather vegetables and bedding plants.

✄ *Water:* Continue to water as needed. Deep-water young trees and new plants before the ground freezes.

✄ *Weed:* Clean up weeds and plant debris.

✄ *Feed:* Add bonemeal when you plant spring bulbs. Feed remaining vegetables and annuals. Do not fertilize roses until spring. Fertilize evergreens, and feed deciduous trees when dormant.

✄ *Cut back:* Cut back chrysanthemums and other herbaceous perennials when they finish blooming. Remove and compost spent annuals and vegetable plants. Wait until winter to prune trees and shrubs.

✄ *Cultivate:* Rake up and compost fallen leaves. As nights grow colder, cover tender vegetables, strawberries, and perennial herbs. Clean and store garden stakes. Dig and cultivate empty garden areas. In colder regions protect roses and other vulnerable plants with winter covering, mulch around trees and shrubs, and stake young trees. In milder regions continue to divide and replant crowded perennials.

❧ **Harvest:** Harvest apples, pears, raspberries, and other fall fruit. Dry and store onions, potatoes, and other root crops. Harvest late fruits and vegetables, and enjoy any tomatoes that have ripened inside. Dry herbs to put away for winter.

❧ **Watch for:** Pick up fallen leaves and fruit from the ground. Continue to check for pests and disease and treat as needed. If you have brought in dahlia tubers and other tender bulbs for the winter, check them for rot or mildew. Also check stored winter squash and root crops. Discard any affected bulbs or vegetables.

❧ **Enjoy:** Enjoy the ripe fruits and golden harvest of Thanksgiving.

# NINE

## September:
## Change of Seasons

While the earth remaineth, seedtime and harvest,
and cold and heat, and summer and winter,
and day and night shall not cease.

GENESIS 8:22
*King James Bible*
(1611)

ON SEPTEMBER 9, Bob and I noticed a change in the weather as we walked to the lake at sunset. The sun had disappeared beneath a layer of clouds, turning the water and trees darker shades of green under an iron gray sky. A streak of lightning flashed across the horizon. Then thunder rumbled in the distance, repeating in more rapid succession as we walked quickly home. That night we had a real thunderstorm, the first rain of the season. Thunder is so rare here that our dachshund Heidi ran around barking at what sounded like some giant animal growling outside our house.

September marks the beginning of autumn. The seasons change, days grow cooler, and night falls more quickly. Each September transforms eastern foliage into symphonies of red and gold, brings the first snowfall to the Rockies, and precipitates a series of changes in our gardens. Summer fruits and flowers disappear, gardeners prepare for winter, and children pack away their summer games to return to school. As much as I love summer, I always feel a new rush of energy in the fall. I love the russet tones of this transition time, the autumn leaves, golden pumpkins, and air as crisp as ripe pippin apples. For me September has always been a season of new beginnings, a time to shed the languor of summer to focus on new, more active goals.

In September the great wheel of time turns inevitably from summer to fall. From the Middle Ages through the Renaissance, people began harvesting their grain at Lammas (August 1), hastening to gather it in by Michaelmas (September 29) or their crop would be spoiled by stormy weather. Our agrarian ancestors measured their time by the seasons, each with its own tasks, feasts, and remembrances. The agricultural year took these people on a spiritual journey through the annual cycle of life, death, and rebirth. Each month held new lessons, as alternating seasons of darkness and light prompted men and women to seek deeper meanings beneath the busy surface of their days.

Gathering in their last crops in September, they enjoyed the feast of harvest home, taking time to celebrate, reflect, and prepare for winter. This was a time for harvest and hiring fairs, where people not only sold their grain and produce, wool, and livestock but also signed new twelve-month contracts as laborers and servants. September 14, Holy Cross Day, was a time for gathering nuts, feasting, and revels. At the autumnal equinox, between September 20 and 23, the weather began to change. An old English proverb for September 21, St. Matthew's Day, said, "St. Matthew brings the cold rain and dew."[1] The seasonal changes were commemorated on September 29, the feast of St. Michael or Michaelmas, which marked the end of the agrarian year when accounts were settled, rents were paid, and local elections were held. People enjoyed the late-blooming Michaelmas daisies and held elaborate celebrations, lighting bonfires and feasting on roast goose. John Milton's mask *Comus* was performed during Michaelmas festivities at Ludlow Castle on September 29, 1634. At Michaelmas people surrounded themselves with warmth and good fellowship, affirming the protection of St. Michael the archangel as they faced the approach of winter.[2]

The wisdom of the seasons has been lost to many in this secular age, when work hours have expanded beyond what even the Puritans could have imagined. But our gardens remind us of seasonal changes, restoring our connection with cyclical time. This month we can share our harvests with friends, enjoy the late blooms and colors of autumn, reflect on the changing seasons, and find our own ways to affirm the light through the coming months of darkness.

## GARDEN GROWTH

IN MY GARDEN September's changes are subtle. Summer flowers and foliage gradually fade, and others take their place. Early this month I noticed some tall cannas blooming near the arbor outside my study window. Cannas (*Canna hedychium*) are late summer bulbs that like sunny locations and regular moisture. I cut back each stem when it finishes flowering so that new stems will blossom throughout the early fall.

These cannas, a beautiful peach color, echo the lighter peach-colored roses blooming nearby. Gardening gives us new opportunities to celebrate color and design. Between the peach-colored blossoms a gray-green Mexican sage bush (*Salvia leucantha*), brought to the ground by last winter's frost, now rises with renewed vigor, raising its abundant spires of purple blossoms.

Not far from the roses, a Chinese lantern (*Abutilon hybridum*) bush is filled with abundant red blossoms, responding to last month's application of fertilizer. The violet rose of Sharon by my window has finished blooming, so I trimmed it back, but now the bush beside it—Brazilian plume flower (*Justicia carnea*)—is filled with clusters of pink flowers. In our gardens, as in our lives, as some blossoms fade others appear. Part of the joy of cultivation is recognizing these changes, finding small moments of beauty all around us.

Promising blooms in late autumn, the chrysanthemums (*Chrysanthemum morifolium*) growing in a large wooden barrel beside the deck are covered with tiny buds. They, too, have responded to last month's repeated applications of fertilizer. I gave them another helping of high-phosphorus 10-60-10 this week, as their buds are about to open.

In my garden September brings a succession of new discoveries. Clusters of small purple berries have appeared on the elder trees (*Sambucus mexicana*) high overhead. Just yesterday I noticed blossoms on another tree—a white crape myrtle—outside my bedroom window.

Along the south yard the tomatoes are still bearing fruit, providing daily harvests for our salads. Beside the north fence the green beans have died back, but in the friendship garden the purple dahlias and gold California poppies are still blooming, along with an occasional tiger flower. Most remarkable, though, is Rhonda's society garlic (*Tulbaghia violacea*), still filled with clusters of violet blossoms. These plants have bloomed for months now, providing color and abundant foliage with very little care. I'd love to plant some in the iris garden to provide a backdrop of color there next summer.

In our front yard the roses are taking turns blooming. Sometimes the red-and-white Double Delights take precedence, sometimes the red Mister Lincolns, yellow Sutter's Gold, or other blossoms in apricot, gold, white, or tangerine. Near the roses are Betty's dahlias, which were blooming beautifully when we moved in last fall. But this year I've had mixed results. One dahlia, a bright tangerine, has grown stronger and bloomed more than all the others, and some have not bloomed at all. Since I've never grown dahlias before, I'm still learning about them, wondering what has made one plant so much more robust. One possibility may be fertilizer. The prolific dahlia, growing beside the yellow rosebush, may have benefited from the rose stakes I put there. Next year I'll fertilize all the dahlias more and see what happens.

In the iris garden the nasturtiums are flourishing in the cooler weather, their abundant red, orange, and yellow blossoms outperforming even the lobelia and alyssum. While weeding and trimming spent bedding plants this week, I found a happy surprise. There, poking up out of the soil, were tiny green shoots—the grasslike leaves of grape hyacinths (*Muscari arme-niacum*). I smiled in recognition. I love these plants with their bright purple clusters that appear in early spring. Every day I see more of them emerging, filling in the bare spots in the iris garden between the herbs and flowers. It's so good to see them again, especially now, when other plants are fading. The grape hyacinths are a gentle reminder of the promise of spring as we head into the dark days of winter. Their reappearance is like a welcome visit from an old friend, an affirmation of nature's eternal cycle.

## GARDENING AS SPIRITUAL PRACTICE

### *Cultivating Order Within and Around Us*

LAST WEEK AFTER a day of meetings at the university, I came home to spend an hour in the front yard picking up leaves, pulling up small weeds and grass shoots, and trimming back yellowing iris leaves. Small actions, they combine to create greater order and beauty.

Later, when I went inside to work in my study, I found myself moving more mindfully around the room, putting things away when I'd finished a project, clearing clutter, cultivating greater order and beauty in this space as well.

What we practice in our gardens extends to the rest of our lives. Care and cultivation: this practice is not some dramatic before-and-after makeover but an ongoing spiritual exercise. Small actions performed on a regular basis are the building blocks of our lives. This month, as you perform the many garden tasks to move from summer into fall, look for ways to extend your garden wisdom into the rest of your life.

### *Maintaining Your Edge*

A FEW DAYS ago I visited the beautiful Elizabeth F. Gamble Garden Center in Palo Alto. Since gardening is a continuing education, visiting nearby botanical gardens and conservatories provides a wealth of new insights and ideas.[3]

Walking through the Gamble grounds, I noticed how attractive the garden beds looked. Blending herbs, vegetables, and flowers, the beds were neatly edged, set off like artwork carefully matted and framed. When I came home I looked with new eyes at my front yard. The cobblestone walk leading from the street to the front gate is surrounded by grass and clover, with garden beds on either side. Earlier this week I'd run the hand mower over the small grassy area but hadn't noticed how much the edges needed work. The grass and clover had grown vigorously, reaching into the driveway and over many of the cobblestones. Sending runners and underground roots into the garden beds, these plants had blurred the edges, invading the homes of herbs and flowers.

To regain my edge I trimmed the grass by the driveway and cobblestones with my bright red grass trimmers. Then I went around the garden bed with my garden knife, cutting a clean new edge and removing the errant grass and clover. As I was doing this, I thought to myself how we much need to check our edges on a regular basis, not only in our gardens but in our lives, so that invasive grasses don't choke out what we're trying to cultivate.

Living in California's Silicon Valley, I hear complaints every day about how much people's work intrudes on their personal lives. Not only are people working longer hours, with less time for contemplation, but all kinds of interruptions—e-mail, faxes, pagers, and cellular phones—are intruding into their private space. Where is the edge between lawn and garden, public and private? In this busy postmodern world, each of us must consciously work to maintain our edge. A few years ago, when I was chair of my department, some colleagues would call me at night to talk about schedules, meetings, and evaluations. After numerous interruptions during dinner, I decided to put my work back where it belonged. I let my answering machine take routine calls, then returned them from the office the next day. Our work is important, but some things in life we cannot delegate: time to eat, sleep, exercise, contemplate, and cultivate our relationships. No one else can do these for us. In order to make time for them, we must maintain our edge.

### Personal Exercise: Taking Action to Maintain Your Edge

HOW WELL DO you maintain your edge? Do you allow time for nurturing meals, rest, exercise, contemplation, and communication with those you love? If you've become "too busy" for these things, something is wrong. Invasive grasses from the outside world are intruding on your life's garden. It's time to take positive action.

🐾 The first step in changing any behavior is awareness. Ask yourself what parts of your life need cultivation. What would you like more of?

🐾 The next step is to take action: Make more time for what you value. Schedule in some time for what you've been missing.

🐾 Now make room for this activity. Ask yourself, What are the invasive grasses in my life, and how can I begin eliminating them?

🐾 If you can't think of anything to cut down on, spend this week taking an informal time inventory. Each day keep a log of how you spend your time. Write this in your daily planner or on an index card. At the end of the week, look over your time log for "invasive grasses," time-wasting activities.

🐾 What do you find?
  · intrusive phone calls?
  · a negative person in your life who monopolizes your time with complaints and calamities?
  · too many evenings spent in front of the television set?
  · aimless shopping?
  · something else?

🐾 Choose one invasive grass you'd like to eliminate and figure out a way to do so. Let an answering machine take your calls, limit your time with the negative person, resist the temptation to turn on the television every night or wander through the shopping malls. Identify one major time waster and eliminate it.

Be patient with yourself and the process. If your life is extremely out of balance, it did not get that way overnight. It will take time to get things back in order, but the mindful action you take now will create positive momentum. You'll feel more rested. You'll also have more time to cultivate the healthy new practice you began last month. If you haven't managed to set up a new walking practice or other healthy habit as you intended, give yourself another chance. Review the pages from last month and schedule time for your chosen practice. This month, with those time wasters out of the way, you'll find the going smoother.

Keep eliminating the invasive grasses that steal time away from what's important to you. Each of us has certain traps and pitfalls at work. If I go back to my office at 5:00 after a meeting, I'll get caught up in trivia—checking my e-mail, going through papers on my desk—and spend another hour or more as my energy drains away. Accomplishing little and exhausted when I leave, I miss my evening margin for exercise, contemplation, and renewal. To avoid this trap, I now take my briefcase and coat to the meeting and, when it ends, go straight to my car, taking myself home without getting trapped.

## Turning toward the Light

WHEN I WATER my houseplants, I rotate them to keep them growing symmetrically because they always lean toward the light. They have a kind of intelligence that moves them toward as much light as possible. Botanists call this *phototropism*: when a plant's stem grows toward the light, leading to asymmetrical growth.

Plants have more than one way of turning toward the light. When they move their leaves to follow the sun, it is called *heliotropism*. Plants usually keep their leaves perpendicular to the sun's rays, so they can absorb as much solar energy as possible to facilitate photosynthesis, which uses light to convert carbon dioxide and water into carbohydrates they use for food. In extreme heat or drought, plants turn their leaves up vertically, to reduce the amount of sun on their surfaces. This is what my bean plants were doing last month until I gave them some afternoon shade.[4]

Plants always turn toward what is good for them, while many people do just the opposite: dwelling on the darkness, lingering in the shadows. Like a garden in filtered sun, any day brings us both sunlight and shadow, good and bad events. Some people focus on the good, turning toward the light. These are the optimists. Others dwell on the dark side of everything. Focusing on the negative, they ruminate and catastrophize. Busy worrying about what might happen, they miss the bright opportunities in their midst. More people than ever seem to be dwelling in the shadows. According to the psychologist Martin Seligman, there's "an unprecedented epidemic of depression" in America and most of the developed world, but we can do something about it.[5] We can change the way we respond to our experience: stop dwelling in the shadows and start turning toward the light.

This week I sprained my back and found myself staring at the shadows. My back hurt, I felt incapacitated, and I couldn't train in aikido, put compost around the garden, or do a lot of other things I'd planned. The sprained back was bad enough, but thinking about all I could not do and worrying about my body falling apart was even worse. I was dwelling in the shadows.

Turning toward the light gives plants access to the sun's vital energy. People become energized when they affirm a sense of agency, renewed power and possibility. Since I couldn't really work in my garden, I took the time to walk through it slowly. This was much better than sitting inside ruminating. Walking around helped my back muscles loosen up, and being in the garden always lifts my spirits. Focusing on what I *could* do instead of what I couldn't made me feel better too. I couldn't bend over to do much garden work, but I could do some light trimming, so I took my garden shears and cut back some errant crape myrtle and camellia branches to create more order along the north garden path. Then I cut some roses to bring inside.

*Personal Exercise: Seeking the Light in Your Own Experience*

IF YOU'VE BEEN dwelling in the shadows, facing a challenge at home or at work, take a cue from the natural wisdom of plants and begin to turn toward the light. First, ask yourself:

☙ *What, for me, is dwelling in the shadows?*
- Ruminating, obsessing about a problem without taking action, or dwelling on a problem with no immediate solution?
- Blaming myself?
- Blaming someone else?
- Worrying about the worst that could happen?
- Complaining about what I cannot do?

Then turn away from these habits and begin to see the situation differently, asking:

☙ *What, for me, is turning toward the light?*
- Focusing on what I *can* do?
- Looking for new possibilities in the situation?
- Finding unforeseen gifts, blessings, opportunities?

When I recently called one of my friends, a dedicated coach and athlete who loves his sport, he said he had gotten an injury riding his mountain bike. But, ever the optimist, he was looking at the bright side. "I couldn't work out, so I've had a chance to spend more time with my family and get caught up on a number of projects," he said, focusing on what he *could* do and enjoying it.

There will always be patterns of sunlight and shadow. What makes the difference is how we respond to them. Attitude is one of the most important things we can cultivate in the gardens of our lives.

## GARDEN TASKS

### Planting Vegetables and Herbs

IN SOME AREAS the garden year is coming to a close. In others it's time for cool-weather vegetables. In zones 1 through 4 you'll need to cover your last crops to protect them from frost. Depending on local weather conditions, you can still sow cool-weather vegetables in zones 7 through 11. In zones 5 through 6, you've probably already begun growing your last crops in a cold frame, while in zones 9 through 11 it may still be so warm you'll need to start your seeds indoors. In my area (zone 9) of Northern California, there's time to sow garlic, peas, spinach, chard, beets, carrots, radishes, parsnips, and other cool-weather crops in the garden. This month I sowed some more beets in the raised bed on the south side of the house. Last month's seedlings were eaten one night—perhaps by snails and slugs, so this time I scattered snail bait around the perimeter of the raised bed.

Today I planted some onion sets by the strawberry plants in the front courtyard. Our entryway now looks very welcoming, with our new bonsai elm by the front door and pots of begonias, basil, and spinach grouped nearly. All of this is not only aesthetic but convenient—it's easy to come out and snip some basil and green onions for dinner or water all these plants with the green-and-gold watering can sitting beside them.

In the front iris garden, among the nasturtiums and alyssum, I set out the pansies I'd started indoors. In moderate climates this is a good time to sow or set out cool-weather flowers (sweet peas, sweet alyssum, and the prolific California poppies, which reseed themselves throughout my front gardens). While the ground temperature is still at least 54 degrees Fahrenheit, they will produce their best growth. I only hope the afternoon sun is not too intense for the new pansies and will watch and water them carefully.

The rosemary plants I set out in the front gardens last year have taken hold and are now growing abundantly, providing fresh herbs for many fall and winter recipes. Rosemary was a favorite Renaissance herb. As Shakespeare tells us in *The Winter's Tale*, it would "keep seeming and savour all the winter long."[6] Originally a Mediterranean herb, rosemary has become so entwined with English tradition, it's hard to discern just when it was first cultivated there. Some say that the early Romans imported it, that the Saxons used it as a healing herb, that it was introduced in the 1360s by Queen Philippa of Hainaut and again in the 1500s by Sir Thomas More, who grew rosemary on his garden walls, as it later graced the royal garden at Hampton Court.

"There's rosemary, that's for remembrance," said Ophelia, distributing her herbs in Shakespeare's *Hamlet*.[7] Associated with love and friendship during the Renaissance, sprigs of rosemary were given out to wedding guests, strewn before the bride, and used to decorate the bridal cup. Houses and churches were decorated with them during the Christmas season, and rosemary bouquets have been used since classical times to strew on graves so that the departed would not be forgotten.

Medieval and Renaissance households were infused with the scent of rosemary. People used the herb to flavor meats and wine, strew on floors and in wardrobes, and burn as an air freshener. They drank rosemary tea as a tonic to calm and cleanse their systems, used rosemary to flavor sugar, wine, and honey, made rosemary candy and cosmetics. The medicinal uses of rosemary are legendary. Gerard's *Herball* says that rosemary flowers blended with sugar would "comfort the heart, and make it merry, quicken the spirits, and make them more lively."[8] Many people wore sprigs of rosemary or sprinkled rosemary powder on their clothes. Rosemary was also believed to comfort the brain and dispel bad humors, so many people put it on their pillows before going to sleep.[9]

Rosemary grows wild in Mediterranean climates, apparently thriving on neglect, so here in Northern California I've always taken it for granted. Since it is my favorite culinary herb, I set small rosemary plants out all over the yard when I moved in last fall. The results were revealing. The rosemary growing in a dry sunny spot by the sidewalk took hold and prospered. The plant growing in the shade underneath the birch trees slowly languished, and the plant growing in the moist earth in the north garden succumbed to root rot and mildew. After that I planted other rosemary plants in sunny parts of the iris and herb gardens.

Plants, like people, have their own needs and preferences. When the thermometer approaches 100 degrees, my friend Cory comments cheerily on the weather while others complain about the heat. Like Cory, rosemary thrives on warmth and sunshine. But cold and damp it cannot abide. If your region has cold winters, rosemary is best kept outdoors in a pot in spring and summer, then brought inside by a sunny window as the seasons change this fall.

New surprises and new growth are still possible this time of year. The basil I picked last month from Rhonda's yard and placed in a jar of water had rooted, so I planted the cuttings in a pot in our front courtyard. If you live in a colder region, now is the time to transplant tender geraniums, coleus, begonias, and many herbs into pots to overwinter inside. You can also grow many herbs from cuttings now to give as holiday gifts.

Thomas Tusser's guide for Renaissance gardeners reminded people in September to pot up their herbs to bring inside for the winter. This was also the time to dry and store herbs to season winter stews and pottages.

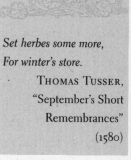

*Set herbes some more,*
*For winter's store.*
THOMAS TUSSER,
"September's Short
Remembrances"
(1580)

## Drying and Storing Herbs

ONE CHRISTMAS SEASON I gave friends and family jars of homegrown herbs, carefully labeled and tied up in red ribbons. It's time to begin drying and storing herbs for winter, both for your own use and for gifts. Here's how:

- Harvest herbs in the morning before the sun has dried their oils. Ideally, pick leafy herbs like basil and thyme before they bloom. Flowering herbs like lavender dry best when harvested as the flowers begin to open. But life is seldom ideal; it's still worth harvesting your herbs at any stage.
- Cut sprigs of herbs with garden shears or scissors, removing the outer leaves from plants like parsley and the tops of rosemary and thyme if you'd like to encourage further growth. If you're harvesting an annual, such as basil, at the end of the season, cut down the entire plant.
- Remove large leaves (such as bay and basil) from the stems. Leave smaller leaves on.
- To dry herbs, either spread them on a screen in a single layer or tie them in bunches and hang them from hooks, a drying rack, or an old coat hanger.
- Place them in a warm, dark place: dark so the sun won't evaporate their oils, dry so the herbs won't mildew.

- Check the herbs and, if laid on a rack, move them around every few days to eliminate any damp spots.
- You can also dry your herbs on a cookie sheet in a gas oven with just the pilot light or an electric oven on the lowest setting. Keep the door slightly open to release moisture. Check the herbs and move them around periodically to eliminate damp spots. Remove the herbs when they are completely dry, like the ones in your spice cabinet. Depending upon the herbs and the heat of your oven, this process will take several hours, filling your home with delightful aromas.
- Remove the dry leaves from the stems of herbs such as rosemary, thyme, and oregano.
- Place the leaves into small airtight jars, label them, and keep in a dark, dry place: a closet, pantry or kitchen cabinet.

### Feeding Your Plants

THIS MONTH YOU'LL need to fertilize any flowers and vegetables still growing in your garden. Feed your chrysanthemums for the last time right before the buds open. When you plant spring bulbs, amend the soil and add bonemeal to help them when they start to grow. If your region experiences frost and cold winter weather, do not fertilize woody plants until next spring.

Since we have a long growing season, this month I've been fertilizing the tomatoes with a water-soluble fertilizer every week. I've also given repeated feedings to the chrysanthemums and Chinese lantern bush in back, the bonsai tree and the mesclun greens seedlings growing in containers in our inner courtyard, and all the annuals in the front gardens.

### Watering, Planting, Pruning, and Weeding

SEPTEMBER CAN OFTEN be quite hot in our region, so I've been watering regularly, keeping an eye on the new plants and seedlings. Fall is a great season to plant new trees, giving them time to get their roots established without the stifling summer heat. But you'll need to make sure new plantings get enough water.

As many plants stop blooming and die back, this is a good time for pruning. In cooler regions you'll need to wait until trees and shrubs are dormant. In the front yard I pruned the lavender, which has finished blooming. With the cooler nights and occasional rain, I've noticed powdery mildew on some of the roses and dahlias, so I've been spraying them with fungicide.

## Caring for Irises

THIS MONTH I'VE gotten to know my irises, clearing away their dried foliage and trimming back their remaining leaves into six-inch fan shapes. I've also been reading about irises in Gerard's *Herball*. This Renaissance classic, first published in 1597, contains many illustrations of irises, known then as "floure de-luce," after the French fleur-de-lis, the yellow iris that became the symbol of France. Gerard says that irises bloom from late March until early May and believed that their roots could heal bruises and strengthen joints.[10] The irises in my garden certainly strengthen my optimism with their remarkable range of color. In September all their faded foliage and rhizomes look alike, but last spring I could hardly believe the parade of colors: not only purple, violet, white, and gold but delicate peach, violet mixed with gold burgundy, eggplant, even pale green.

Cleaning up the iris foliage has led to much-needed weed and pest control. The other day I took out at least two dozen dandelions that had been hiding among the faded foliage. I also saw snails on some of the leaves, so I put out more snail bait. All the discarded iris foliage went into the compost, as did the clippings when I sheared the spent blossoms from the alyssum and California poppies.

## Dividing Irises

SEPTEMBER IS A good time to divide your irises if you have not already done so. If any in your garden failed to bloom last spring, they're probably overcrowded or lack nourishment. You can promote more blooms next spring by dividing the plants, amending the soil, and adding plenty of bonemeal. Usually irises need dividing every three to four years. Overcrowded irises will have a clump of rhizomes growing in a circle around an old, spent rhizome in the middle. Only the plants on the outer edges will bloom. The cluster of irises in the back garden did not amount to anything last spring, so I decide to divide them. First I assembled all my tools:

- sharp knife
- garden shears
- garden fork
- spade
- sulfur-based fungicide to treat the cut rhizomes
- a bag of compost to amend the soil
- bonemeal to nourish the plants

Then I began carefully digging up the rhizomes. Since the soil was hard and dry (clay soil that desperately needed amending), I first loosened it with my garden fork. The next day I dampened the soil and carefully dug up the clump of rhizomes, lifting them out of the soil and setting them aside. Using my spade, I dug into the soil. It was as hard as cement, so I added plenty of compost to improve the texture.

Going back to the plants, I washed the soil off their roots and inspected them closely. A fascinating part of gardening is literally getting to the root of how things grow. This clump of irises had many rhizomes growing from an old central core. I cut off the outer rhizomes with my knife, leaving each plant with a root longer than my index finger. Then I cut away any parts that looked rotten and dusted the cut surfaces with fungicide to prevent disease. I planted the newly independent plants in shallow V-shaped trenches, each one fortified with a dose of bonemeal. I placed the rhizome on top and the small roots in the two sides of the V, covering them with soil but leaving the top of the rhizome above the soil surface. I then discarded the old leaves and core rhizome. With enough room to grow and their soil amended, these irises should bloom beautifully next spring.

## Dividing Spring Bulbs

LIKE IRISES, MANY bulbs and perennials can benefit from division and replanting when

* they're crowded
* they're not performing well
* or you wish to move or multiply them

Don Ellis, resident horticulturist at the Gamble Garden Center in Palo Alto, California, says, "Let the plant indicate when it needs to be divided." If it's blooming and performing well, leave it alone, but if it blooms only on the outer edges or not at all, it's time to divide it. Different plants need different division techniques.

*For agapanthus and daylilies:*

* First dig up the plant, being careful not to cut through too many roots, and remove the root clump from the ground.
* Then divide the roots: Insert two garden forks in the middle of the clump and gradually pull them apart, separating the clump into two sections. You can divide smaller clumps with a hand fork.
* Replant the two separate plants in newly cultivated and amended soil.

*For bulbs such as daffodils, tulips, and muscari:*

* Lift the clump of bulbs out of the ground when their foliage has died back and they are completely dormant.
* Untangling the roots, separate the clump by hand, first into smaller clumps, then individual bulbs.
* Check the bulbs for signs of rot, and discard any diseased bulbs.
* Clean the soil and tunics (the onionlike outer skins) from the bulbs. Separate the smaller offshoots from the larger bulbs.
* Replant the bulbs and offshoots separately in newly cultivated and amended soil.[11]

## Harvesting Your Crops

### Drying Tomatoes to Store for Winter

*To dry tomatoes in your oven:*

- *Slice the tomatoes, lay them on a baking sheet, and sprinkle salt over them.*
- *Put them in a 200-degree oven for 5 to 6 hours.*
- *Check the tomatoes periodically to make sure you remove them in time—you want them dried but not burnt.*
- *Let the tomatoes cool, then carefully remove them from the tray, set them into a colander, and rinse them with cider vinegar to remove the salt.*
- *Put the dried tomatoes into jars of olive oil and store them in your refrigerator.*

AS WE ENJOY the final fruits and vegetables of the season, we participate in harvest rituals that date back centuries. In old England, Michaelmas harvests and feasts offered a last taste of summer's abundance before the advent of winter. This month I've been appreciating all the homegrown tomatoes, knowing they won't be around much longer. Every day when I water my tomato vines I pop one or two Sweet 100s into my mouth. These are by far the best performers, outdistancing the cherry tomatoes and Early Girls in their profusion, flavor, and productivity.

As the nights grow cooler and the first frost approaches, there are ways to extend your tomato harvests. You can shield your plants with protective coverings to keep them warmer. You can also dry some of the fruits to preserve them. When frosts are inevitable, pick your green tomatoes and bring them inside to ripen.

My friend Ann told me how she dries tomatoes in her oven and puts them into jars of olive oil, using them like sun-dried tomatoes in sauces and stews. Since I love sun-dried tomatoes and we've been harvesting more ripe tomatoes than we can eat, I tried Ann's recipe, putting up several jars of dried tomatoes to enjoy this winter.

For years my neighbor Rhonda has enjoyed homegrown tomatoes during the winter months by taking green ones inside at the end of the season and placing them in single layers on sheets of newspaper, where they gradually ripen. Not as sweet as those ripened on the vine, but they still taste homegrown.

As much as I love tomatoes and do all I can to

extend their harvest, I'm amazed at the waste of good fruit by generations of English gardeners. Even the authoritative John Gerard explained in his *Herball* (1597) how he carefully planted tomatoes from seeds, growing them as ornamentals in his London garden but never tasting their succulent fruit. He noted that people in Mediterranean countries ate them "prepared and boiled with pepper, salt, and oile." But he believed tomatoes were unhealthy: "They yeeld very little nourishment to the bodie, and the same nought and corrupt."[13] Unaware of their healthy vitamin C and lycopene, poor Gerard and his contemporaries threw their tomatoes away. But to put things in perspective, in his day people still believed our bodies were composed of four humors and the earth was the center of the universe.

- *When you want to use the tomatoes, just leave a jar out on the counter for a few minutes to let the olive oil soften. The tomatoes are delicious in a variety of dishes and taste sun-dried.*[12]

Medieval and Renaissance gardeners were wiser when it came to apples, which they harvested from July to September, wrapping the fruits in straw to store for the winter. Apples were grown in England by the early Celts, Saxons, and Romans. Medieval monks loved apple cider and planted apple orchards in their monasteries.

Late September, or Michaelmas, was traditionally the time to harvest apples. Thomas Tusser warned gardeners not to pick the fruit too green or bruise it when they took it off the tree.

From the Middle Ages through the Renaissance, English men and women enjoyed baked apples, apple cider, apple sauces and pies, and prepared apples for many medicinal uses. In the Middle Ages apples were believed to be good for the heart, and Gerard's *Herball* says that they were used externally to heal inflammations and smooth the skin. Taken internally, they were believed to cure melancholy or depression.[14]

In our household a freshly baked apple pie lifts everyone's spirits. An even easier way to enjoy apples is to slice them, sprinkle them with cinnamon, and bake them in the oven or microwave to serve as a side dish with a variety of meals. When our neighbors went on vacation this month, they asked us to watch their garden and pick any ripe apples from their tree.

*The Moone in the wane,*

*gather fruit for to last,*

*But winter fruit gather*

*when Mikel is past;*

*...*

*Fruit gathred too timely*

*will taste of the wood,*

*Will shrink and be bitter,*

*and seldome proove*

*good:*

*So fruit that is shaken, or*

*beat off a tree,*

*With bruising in falling,*

*soone faultie wil bee.*

THOMAS TUSSER,
*Five Hundred Points of*
*Good Husbandry*
(1580)

The small, sweet apples were wonderful baked, filling our house with delightful aromas that recalled many holiday feasts.

### September Harvest Meal: Roasted Vegetables

YOU CAN CELEBRATE your own harvest feast by gathering in your garden vegetables and roasting them with olive oil and rosemary. A favorite meal of mine is roasted red potatoes, carrots, onions, bell peppers, and mushrooms, which can be eaten as a vegetarian dish or cooked along with roast turkey or chicken breasts.

The cooking times will vary depending on the sizes and types of vegetables, but here are some guidelines:

• *Oven-roasted new potatoes.* Cut potatoes in half (or quarter large ones), and place on a cookie sheet or in a shallow pan sprayed with cooking oil. Drizzle with olive oil, and scatter fresh rosemary on top. Roast in a 400-degree oven for 30 to 35 minutes.

• *Oven-roasted carrots.* Cut carrots in one-inch chunks and roast the same way as new potatoes.

• *Oven-roasted bell peppers, onions, and mushrooms.* Cut peppers and onions into finger-sized slices, remove stems from mushrooms, and cook the same way as potatoes only for 10 to 15 minutes.

You can roast the following along with your vegetables:

• *Oven-roasted apples.* Cut apples into slices and place them on a cookie sheet along with the vegetables. Sprinkle apple slices with cinnamon and bake for 10 to 15 minutes.

• *Oven-roasted chicken or turkey breasts.* For a roasted meal with a meat entrée, put turkey or chicken breasts in a roasting pan, drizzle with olive oil, and scatter rosemary sprigs on top. Bake at 400 degrees for 20 to 30 minutes, depending on the size of the poultry, then add the vegetables and cook for the times indicated in the previous lists. The meat and vegetables will cook together till they are finished (check poultry for doneness with a meat thermometer). In about an hour you'll have a complete roast dinner.

The aroma of rosemary will gradually fill your house, turning the vegetables into a simple, savory feast. A roast dinner nearly cooks itself, freeing up time for you to relax and enjoy the colors of this transitional season.

## GARDEN REFLECTION

### *Cultivating Resourcefulness*

GARDENING CAN CULTIVATE resourcefulness. I've seen gardeners use gallon plastic bottles with the bottoms cut out to protect their plants from frost this time of year or grow seedlings in cut-down milk cartons. They save time and money by using the materials around them.

This kind of thinking has been going on for centuries. In 1580 Thomas Tusser advised Renaissance gardeners to collect troublesome stones from their gardens and use them to build walls and pathways. That's exactly what the award-winning gardener Peter Chan did at his house in Oregon. When he moved in the yard was littered with weeds and debris, and the hard clay soil was filled with rocks. A few years later, after Peter's patient work, his soil was amended and the stones had been laid in neat pathways around raised garden beds. His beautiful gardens have been featured in national magazines and on network television.[15]

*When stones be too manie, annoieng thy land,*
*Make servant come home with a stone in his hand.*
*By daily so dooing, have plentie yee shall,*
*Both handsome for paving and good for a wall.*
        THOMAS TUSSER,
    *Five Hundred Points of Good Husbandry*
        (1580)

When I married Bob and moved into his house in San Jose, the back-yard garden plot was overgrown with weeds and the sun-baked clay soil was like mission adobe. Planting a vegetable garden there took time and resourcefulness. I soaked the soil for a couple of days to make weeding easier, then spent the next few days pulling the weeds and grasses. When I turned over the soil, I saw how thickly entrenched the grass roots were, so I tried an experiment.

We had stacks of newspaper in the garage, ready for recycling. But newspapers can also be used as mulch, so I began spreading thick layers all over the garden, wet them down, then covered them with a layer of potting soil and compost. When I set out the vegetable plants, I made small holes in the newspaper layer with my trowel. I sowed the seeds in the compost-soil mixture on top.

Since Bob was born in New York, we subscribe to *The New York Times* on Sundays. We began joking about our *Times* vegetable garden: "All the news that's fit to eat." But the papers did an admirable job of keeping down the weeds. That year we had corn, yellow squash, carrots for Bob's horse, spinach, basil, thyme, oregano, parsley, onions, and tomatoes. The next year in place of the corn we grew four rows of green beans. By then the paper had begun to break down into the soil while still keeping away most of the grasses and weeds. When we sold Bob's house last July, the garden was filled with an abundance of vegetables. We were still harvesting them when we moved out last September.

This month, as seasonal changes bring their inevitable challenges, remember to follow your garden wisdom. Cultivate resourcefulness, using the materials around you to create new possibilities in your life.

# TEN

## October: Autumn Harvests and Perseverance

For his bounty,
There was no winter in't; an autumn 'twas
That grew the more by reaping.

WILLIAM SHAKESPEARE,
*Antony and Cleopatra*
V.2, ll. 85–87 (c. 1607)

OCTOBER BRINGS A mosaic of changing colors. Red and orange nasturtiums create luminous patterns in our front yard. Throughout the garden new bulbs raise their tentative green foliage. Birch leaves are scattered across the cobblestone walk like pieces of autumn gold.

In October, as the trees on our street begin turning from green to gold, pumpkins start appearing on the doorsteps. At the end of the month neighborhood children are transformed into goblins, elves, and heroes of their imaginations. Their Halloween rituals date back to the Celtic feast of Samhain, the beginning of winter darkness. In the Middle Ages, Samhain became All Souls' Day and All Hallows' Eve, celebrated with candles, bonfires, roasted nuts and apples. This was a time for games, love charms, stories, and fortune-telling. Groups of "soulers" would go from house to house, representing the souls of the departed, singing and begging for "soul cakes."[1]

Like the neighborhood children, our gardens, too, are transformed in October. As the days grow shorter and nature's colors change to crimson, bronze, and gold, we gather in our late harvests and prepare for winter. In Renaissance England, country churches were decorated early this month with pumpkins, carrots, apples, potatoes, and Michaelmas daisies. Children would bring newly-harvested vegetables to the altar as gifts, which were blessed and given to the poor. In mid-October, livestock fairs and celebrations were held in the English midlands. Fishing villages along the coast blessed their boats and nets. October 28, the Day of St. Simon and St. Jude, was known for its stormy weather.[2]

Throughout North America, the days are growing colder. In New England and the Midwest, gardeners hurry to gather in their crops before the first killing frost. Yet despite the growing cold and darkness, October has its own kind of beauty, offering a blaze of fall color before the winter darkness.

## GARDEN GROWTH

IN MY GARDEN, October brings the last breath of summer. The weather has been cooler at night but we've had some sunny afternoons in the high seventies and low eighties. Thriving in the cool October weather, nasturtiums bloom throughout the front gardens. In the Middle Ages their red and gold blossoms were considered aphrodisiacs.[3] Like the nasturtiums, the roses offer a final fall display in crimson and gold, as well as vanilla, peach, and mandarin orange. The pansies I set out last month are still not blooming, but purple primroses have sprung up by our front gate, and the tomato vines along the south fence are still producing cherry tomatoes and Sweet 100s. Near the end of the month, an early paperwhite narcissus raised

*O Primroses! Let this day be*

*A resurrection unto ye.*

ROBERT HERRICK,
"Upon Julia's Recovery,"
ll. 5–6 (1648)

its tiny blossoms in the friendship garden, where Rhonda's society garlic blooms on, along with the last few tiger flowers.

In our backyard the two standard rosebushes offer their final peach and pink buds, the canna has produced another stalk of peach-colored blossoms, and the salvia is crowned with dozens of purple velvet spires. The red Chinese lantern is still blooming, but the hydrangea blossoms nearby have begun to fade from white to green.

Along the north yard are signs of seasons to come. Early spring bulbs are raising their foliage, camellias are forming buds, and rows of red berries deck the cotoneaster (*Cotoneaster horizontalis*) for the holidays.

For the past few weeks I've been watching the chrysanthemums, anticipating their late-season burst of color. On October 7 the buds were nearly open, promising blossoms of burgundy and rust. Three days later the first blossoms appeared: a group of single, two-inch daisy chrysanthemums with rust red petals and gold centers, along with larger decorative chrysanthemums in pale rose. During the next few days other blossoms followed: small lilac pompons, lacy spider chrysanthemums in white and plum, and more decorative blossoms in gold, pumpkin, burgundy, and rose. The result was a vivid autumn display cascading over the sides of its wooden barrel, providing a vision of beauty out my study window and autumn bouquets to grace our dinner table. I have a special feeling for these late bloomers, whose blossoms last into November. While most poets celebrate the beauties of spring, wisdom uncovers the beauty in every season.

*Faire pledges of a fruitfull Tree,*
*Why do yee fall so fast?*
*Your date is not so past;*
*But you may stay yet here a while,*
*To blush and gently smile;*
*And go at last.*
ROBERT HERRICK,
"To Blossoms,"
ll. 1–6 (1648)

*No spring nor summer beauty hath such grace*
*As I have seen in one autumnal face.*
JOHN DONNE
"Elegy IX: The Autumnal,"
ll. 1–2 (1633)

## GARDENING AS SPIRITUAL PRACTICE

### Pushing Through Resistance

NOT LONG AGO I noticed some sweet alyssum growing through a crack in the sidewalk in front of our house. You've probably seen wildflowers growing in unexpected places: reclaiming vacant lots or pushing their way through concrete.

Without all the pampering we give to our gardens, these hardy, independent plants germinate and grow in deprivation, with substandard soil and little water. Yet they survive.

Last night Bob and I talked about perseverance. "Lots of our students haven't learned to push through resistance," Bob said. "They try something and if they can't master it right away, they quit and go on to something else. They don't work through it, push through resistance."

There are reasons for this. How can young people learn perseverance when our world doesn't reinforce them for it? Indeed, how can any of us have faith in our ability to achieve our goals? We live in an environment of instant gratification: fast food, microwaves, instant entertainment on the television or Internet, purchases with no money down. Raised in a culture that emphasizes products over process, people expect immediate results. If we try something and don't succeed right away, too many of us cave in, doubting ourselves, convinced we can't do it—whatever "it" happens to be.

Our gardens affirm another paradigm: the wisdom of process. In time a seed develops into a plant, blossoms and bears fruit. But there are no instant harvests. I can order fresh produce any morning on my computer and have it delivered by Webvan that afternoon. The fresh tomatoes in my garden took much longer and, to me, are far more valuable. Not only are they tastier and healthier but they've taught me perseverance.

> If little labour, little are
>    our gaines:
> Man's fortunes are
>    according to his paines.
> ROBERT HERRICK,
> "No Paines, no Gaines"
>    (1648)

Like the vegetables from our gardens, strength of character does not come instantly. It results from many small actions: successes, failures, obstacles overcome, and lessons learned. This process of perseverance cultivates the gardens within and around us, creating external success and inner strength.

This week I was talking to my student Linda, an aspiring writer, encouraging her to send out her poems and stories. "I don't like other people to judge my writing," she said. "I sent one story to a contest last year, but it came back with a rejection letter."

"What did you do then?" I asked.

"Oh, nothing," she said. "I just put it away."

I know that feeling well. Years ago, as a young professor, I was frustrated trying to publish my first articles. Whenever one was rejected by an academic journal, I'd put it in my bottom drawer, convinced that it was not "good enough." One article sat in my drawer for five years. Then one day I sent it out and it was published in three months.

Sometimes a good article has just been sent to the wrong journal: a seed planted in the wrong place. Sometimes the article needs revision. But if it just sits in a bottom drawer, it will go nowhere.

Any published writer learns to push through resistance. Successful people in many fields—athletics, business, sales, entertainment—learn to *expect* resistance. Athletes won't always win, business colleagues won't always agree with us, customers won't always buy our products, and audiences won't always appreciate our performances. Even in our personal lives, our relationships won't always work out. I recall one young couple who had dated all through college. When they broke up during their senior year, the woman was so devastated she dropped out of school, giving up on her dream of a college education.

In our personal and professional lives, we won't always win, but we shouldn't take resistance personally. Resistance does *not* mean we are worthless or lack what it takes to succeed. Resistance is simply resistance: a fact of nature like the wind or the tides. If we learn to see resistance as part of life, we won't be defeated by it.

## Personal Exercise: Confronting Resistance

HOW WELL DO you handle resistance? Ask yourself the following questions:

- Have you experienced resistance lately
  - in persevering with the healthy new habit you chose this summer?
  - in your work?
  - in your personal life?
- If so, what can you learn from it?
  - Do you need to modify your behavior?
  - Get help or advice?
  - Or just keep practicing?

Now consider how you can keep moving forward:

- Make an action plan for yourself.
- Break it down into manageable steps.
- Take the first step.

Learning to push through resistance is a character strength modeled in our gardens and mastered by successful people throughout time. Like flowers that push through city sidewalks, we can practice perseverance, living with greater confidence, success, and vitality.

## Gathering in Your Personal Harvest

THIS MONTH, AS we harvest our vegetables, herbs, and flowers, we can consider the corresponding harvest in our lives. If you've been persevering with the lessons of last month, maintaining your edge and keeping up with your healthy habit, you are already harvesting greater health and peace of mind.

All of us have many gardens in our lives. Some areas we cultivate: dig, plant, fertilize with copious time and effort. Others we forget or allow to lie fallow. In our fast-paced society most people cultivate their work gardens assiduously, but they often neglect their private gardens: family, friendships, home lives, or creativity. As in our yards, so in our lives. There will be little harvest from the gardens we neglect.

### Personal Exercise: What is Your Personal Harvest?

THIS MONTH SET aside an hour of private time to consider these questions:

- *Your work garden.* Does your work bring you the harvest you desire?
  - Does it provide a meaningful challenge, a creative outlet for your talents, or only pay the bills?
  - What parts of your work bring you a harvest of joy?
  - What parts could be amended and improved?
  - If this area has become overgrown, how can you prune and manage your work to make room for other priorities?
- *Your home life garden.* Do you *have* a home life? Home is our haven for personal nurturing and renewal, but one of the greatest challenges today is the work-life balance. For Dave, a young Silicon Valley entrepreneur, homelife was virtually nonexistent. Until last year he literally lived at work, sleeping on the floor of his office, eating take-out meals, and showering at the local health club. After making millions at an Internet start-up, he finally decided to buy a house.[4]

  Your work-life imbalance may not be this extreme, but if you've been depriving yourself of a nurturing home life, look for ways to regain your balance.
  - Give yourself more time at home. Leave work at a reasonable hour. Perhaps you can telecommute one or two days a week.
  - Begin giving yourself more of the comforts you knew as a child. What represents comfort for you: an extra pillow on the bed? flannel sheets? your favorite food? fresh flowers on the table? your favorite music?

  Adding more time and comfort to your home will renew your joy in life.

🌱 *Your family life garden*. Home life represents your personal nurturing, while family life involves your primary relationships. Nurturing relationships means more than household maintenance, laundry, and meals. Relationships are living, growing organisms that require regular cultivation.

If you've become so busy you've lost touch with your family, what can you do to restore the balance? Find one way this week to cultivate your family life. It might mean planning a family outing together. Perhaps it's as simple as taking more time to listen.

Progressive corporations are adopting family-friendly policies that respect employees as whole persons, supporting their family commitments and work-life balance. If you work for a large corporation, does it provide:

- comprehensive benefits?
- on-site child care?
- flexible hours?
- telecommuting?
- a work-life task force?

Check with your human resources department and suggest improvements. The best way to begin leading a more balanced life is to figure out what you need and ask for it.

🌱 *Your friendship garden*. Our friends give us vital gifts of encouragement and support. Through all the changes and challenges in our lives, there's nothing like the wisdom, compassion, and humor of a friend. But do you take time to cultivate your friendships? If there's someone you haven't seen for far too long, this month get back in touch with a call, letter, or e-mail.

🌱 *Your creativity garden*. We don't have to be poets or artists to live more creatively. Abraham Maslow knew the powerful role creativity plays in self-actualization. By nurturing your initiative, courage, and resourcefulness, a creative project can brighten the rest of your life.

Each of us needs a creative practice, something we do for the joy of it, playing with color and texture, creating something new from the materials around us. Our creative practice may produce music, paintings, furniture, or tapestries, but the essential part is not the product but the process, which engages the right side of the brain, opening us up to valuable insights and inspiration.

## Cultivating Your Creativity

SOME PEOPLE CALL their creative pursuit a hobby; others are embarrassed to admit what they do. Yet throughout the ages many accomplished people have cultivated a creative practice. Winston Churchill painted landscapes, the actress Mary Martin did needlepoint, and former president Jimmy Carter has always found peace of mind working in his wood shop.

Look around at the accomplished people you know. One professor in my department weaves colorful tapestries, another is a master woodworker, and a third is a landscape artist. Across campus another professor is a jazz musician. My aikido instructor does calligraphy, and our former university president writes poetry.

For many of us gardening is a creative practice. We love designing our gardens, cultivating the soil, planting seeds, working with nature, and watching things grow. But this month, as the days grow shorter and winter draws near, begin cultivating a creative practice to take you through the winter season.

- Reach back into your childhood: was there an art or craft you particularly enjoyed?
- Take out that old musical instrument in your closet.
- Go to an art supply store for a sketch pad and pencils or paints.
- Visit a sewing shop for colorful tapestry yarns.
- Go to a hardware store and ask for supplies to refinish an old cabinet or chair.

Choose a creative practice and cultivate it, a little at a time, watching its harvest enrich your life.

## GARDEN TASKS

### Planting Cool-Weather Vegetables and Spring Bulbs

OCTOBER GARDEN TASKS vary dramatically. In milder regions there's still time to sow cool-weather vegetables, while in New England and the upper Midwest people are preparing their gardens for winter. In zones 1 through 4 gardeners have finished their final harvest and are laying winter mulch around the garden. In zones 5 through 7, they are gathering in their last harvests and covering their final crops to protect them from frost. In my area (zone 9) of Northern California, as well as in zones 8 through 11, we can sow garlic, peas, spinach, chard, beets, carrots, radishes, parsnips, and other cool-weather crops, although in zones 8 and 9 they may need protection. This month I sowed beets and spinach in planters. Beets were a popular vegetable in the Middle Ages, believed to relieve headaches and toothaches, while spinach was eaten to strengthen the throat, lungs, and stomach.[5]

### Growing Greens in Containers

IN MILDER CLIMATES this is a good time to grow spinach, lettuce, and beet greens in containers. Here's how:

- Buy a half barrel or large tub at the garden supply store.
- Drill three or four drainage holes in the bottom and fill the container with potting soil.
- Place the container in an area with full sun.
- Add a timed-release fertilizer in the amount specified on the package.
- Plant seeds about an inch apart and cover them lightly with soil.
- Water thoroughly and keep the soil moist.
- Thin the seedlings when they are two to three inches high to about three inches apart. (Eat the thinnings in salads.)
- Feed your greens every week with a water-soluble fertilizer.
- When the greens are large enough to harvest (in forty-five to sixty days), carefully remove outer leaves, leaving the rest of the plants intact. This

way you can make successive harvests as the plants continue to grow.

- Enjoy your lettuce, spinach, or beet greens in salads. Spinach is also delicious steamed for five minutes in a pan sprayed with olive oil and sprinkled with raisins or chopped, dried dates.

This month I planted some organic garlic in a small bed in the back-yard, putting lots of compost on top and timed-release fertilizer in the soil with the garlic cloves. In the Middle Ages people believed garlic improved circulation, relieved flatulence, had antiseptic qualities, promoted overall health, and was an aphrodisiac.[6] Modern research has confirmed many of these beliefs, revealing that garlic lowers cholesterol and blood pressure, relieves gas, and has powerful antiseptic and anticarcinogenic qualities. This useful bulb provides abundant health benefits and is a favorite in many regional cuisines. It's also relatively easy to grow.

### Growing Your Own Fresh Garlic

YOU CAN PLANT your own garlic now to harvest next summer. Here's how:

- Choose a garden area that gets at least six hours of sun a day.
- Amend the soil with compost to improve texture and drainage, or grow your garlic in raised beds. Add timed-release fertilizer if you wish.
- Divide a garlic bulb into separate cloves, being careful not to bruise or break them.
- Plant each clove with the flat end down one to two inches deep, five to six inches apart.
- Apply six to eight inches of mulch over the soil.
- Keep the soil moist.
- In mild climates the garlic cloves will send up shoots in a few weeks. In colder climates, if early foliage is killed by winter freezes, new foliage will appear with spring bulbs.
- In March fertilize with fish emulsion or another water-soluble fertilizer. Fertilize every month or as stated on the package.

• Snip the green leaves as needed for cooking, and harvest the mature bulbs in summer when the foliage dies down. Brush off the soil and use the bulbs fresh or hang them to cure in a dark, dry place about 65 to 75 degrees.

In milder regions October is a good time to plant deciduous trees. If you plant a crape myrtle now, you can enjoy its colorful fall foliage as well as its abundant blossoms next summer. This is also a good time to plant perennials in zones 7 through 11. Candytuft, delphiniums, hollyhocks, and gaillardia will take root now and be ready to bloom next spring. You can also plant spring bulbs as long as the ground isn't frozen. Some bulbs (especially tulips and hyacinths) need a period of cold weather to develop stronger roots. In areas with milder winters you'll want to refrigerate these bulbs now for six to eight weeks and plant them in December. (Keep them in a separate bag away from fruits and vegetables.)

*And this for comfort thou*
*must know,*
*Times that are ill wo'nt*
*still be so.*
*Clouds will not ever*
*powre down raine;*
*A sullen day will cleere*
*againe.*
ROBERT HERRICK,
"Good precepts, or
counsell,"
ll. 5–8 (1648)

This month a small brown box of Holland bulbs arrived in the mail. I put it in a cool corner of the garage until the weekend. Before planting the bulbs in our front yard, I checked their descriptions: eight *Liatris spicata*, which grow into eighteen- to thirty-inch spires of violet and white in late summer; a mixture of fifty *Allium moly* and *Allium sphaerocephalum*, which will produce yellow and red-violet blossoms nine to twenty-four inches high from May through late summer; and twenty-five *Crocus vernus*, early bloomers that raise their heads in cheerful lilac, blue, yellow, and white blossoms when the rest of the world thinks it's still winter.

I dug a hole for each bulb with my spade, amended the hard clay soil with plenty of compost, put in some special Dutch bulb fertilizer, and planted the bulbs twice as deep as they were high. Since the liatris are the tallest, I planted them first, in front of the first row of irises, to provide a row of flowering spires in summer when the iris foliage is dying back.

The next day I planted the allium bulbs in front of the liatris, to provide a backdrop of color throughout the summer. A couple of days later I planted the crocuses at the front edge of the iris and herb gardens, near our cobblestone walk, where we'll discover their early blossoms one morning as winter turns to spring.

The next day I walked out into the front yard trying to imagine what all the new plants will look like. Bulbs are wonderful secrets, buried treasure. The only evidence of them now is the newly cultivated soil. Timed deposits in our garden banks, they will provide beautiful interest for many seasons to come.

## Feeding Your Plants

IF YOU PLANT spring bulbs this month, remember to add bulb fertilizer or bonemeal when you amend the soil. You'll also want to fertilize the flowers and vegetables still growing in your garden and give your chrysanthemums a final dose of fertilizer before the buds open. This month I fed the tomatoes growing on the south side, the chrysanthemums in back, the nasturtiums and annuals out front, the bonsai elm, and all the plants in containers.

## Watering, Weeding, and Tending

DEPENDING ON THE rainfall in your region, you may still need to do some watering this month. I've been watering our tomatoes daily, since they're growing in containers and our south yard still gets quite warm.

These sunny fall days are good times to walk around and check your garden. Some areas will need weeding. I patrol the iris beds and pull out dandelions with my Japanese knife. The rainstorms this month have left the ground damp enough for yellow oxalis to spring up throughout the front gardens, so I've been

*We blame, nay we despise*
*her paines*
*That wets her Garden*
*when it raines:*
*But when the drought has*
*dri'd the knot;*
*Then let her use the*
*watring pot.*
*We pray for showers (at*
*our need)*
*To drench, but not to*
*drown our seed.*
ROBERT HERRICK,
"Out of Time, out of
Tune,"
ll. 1–6 (1648)

pulling out these prolific plants. Some of our plants will need protection from the autumn winds. The other day I tied up the salvia with stakes and green gardening tape. My reward the next morning was seeing a tiny hummingbird darting in among the purple velvet spires as they stood tall once more. When your plants stop blooming and die down, cut them back and compost the healthy stems and leaves. I carry a bushel basket along with me on some of my rounds, collecting deadheaded roses, weeds, and plant debris.

## Watching for Pests and Disease

ONE DAY THIS month I found the lettuce planter in our front courtyard dug up and most of the seedlings gone. "What could have done this?" I asked Bob.

"Squirrels," he said.

"Squirrels? Do squirrels eat lettuce?"

"Squirrels will eat anything," he said with a smile.

So I watched and, sure enough, I saw two squirrels scurrying around our yard, looking for winter provisions. I replanted the planter with beet seeds and put a screen on top. Bob laughed, but the seedlings seem to be growing undisturbed.

## Pest Control Tips

WE ALL HAVE garden pests: sometimes endearing animals—squirrels, deer, and rabbits, sometimes insects, snails, and slugs. If something's been eating your flowers or vegetables, it's time to take action. First, some detective work is in order: What kind of pest is it? Look for signs—maybe the culprit itself is still around.

- Many insect pests blend in with the leaves. Tomato hornworms? White-flies? Aphids?
- Pick larger pests (hornworms and snails) off by hand and discard them. (I use gloves for this chore.)
- For smaller insects use the least invasive control to protect yourself, your

garden, and the environment. I prefer insecticidal soap for small insects and a new organic product, Sluggo, for snails and slugs.

- You can prevent slugs and snails as well as disease by keeping your garden clean, picking up fallen fruits, blossoms, and leaves where these pests hide.
- Perhaps your pest is not an insect but a hungry mammal. Squirrels, mice, deer, and rabbits can wreak havoc in a garden. The best step here is prevention. If squirrels and mice are a problem, protect your planters with screens, or line your tulip beds with hardware cloth to keep these small rodents from eating them (they won't eat daffodils and narcissus). Keep rabbits out of your garden with a two-and-a-half to three-foot-high fence of chicken wire that extends at least six inches underground (rabbits dig). Deer are more difficult. They eat all kinds of plants and flowers and can leap high fences. I've seen people enclose entire bushes with chicken wire to keep deer away. There are a number of deer repellents on the market. Check with your local garden supply store to see what's been successful in your area.

As the weather changes this month, your plants could be more susceptible to disease. Today I found powdery mildew covering the ajuga in the inner courtyard. I sprayed fungicide all over them, filling the courtyard with the pungent smell of sulfur (fortunately, we weren't expecting guests). I then reset the front watering system to ten minutes instead of twenty. The nights are much cooler now, and the weather report predicts rain.

### Harvesting Your Crops

OCTOBER WAS LATE harvest season in the Renaissance and Middle Ages, as people gathered in their final crops to prepare for winter. Apples were made into cider and pears into perry, fermented pear juice. Grapes were made into wine and verjuice, a kind of vinegar used for cooking. The English made their own wine until the climate changed in the fourteenth century, bringing cooler, wetter summers. By the Renaissance, most wine was imported from the Continent and English grapes made into verjuice.[7]

The cooler October weather beckons us to pick our last summer crops and store some away for winter. In my garden the larger tomatoes are gone now, but we still have some cherry tomatoes and Sweet 100s. Last week I harvested and dried three trays of tomatoes to flavor winter soups, sauces, and vegetable dishes.

Since annual herbs won't last much longer, now is the time to harvest and put away your last crop. I picked a basketful of basil, added some dried tomatoes, and made a large batch of pesto, which I froze in small containers to enjoy this winter.

## TOMATO BASIL PESTO

*2 to 2½ cups fresh basil (about one colander full)*
*¾ cup fresh parsley*
*1 cup olive oil*
*⅓ cup pine nuts*
*⅓ cup dried tomatoes*
*3 cloves garlic*
*½ cup grated Parmesan cheese*
*Pinch of salt*

Harvest and wash the basil and parsley. Remove the basil stems and discard. Drain the remaining basil leaves and parsley in a colander.

Combine the olive oil, pine nuts, dried tomatoes, and garlic cloves in a blender. Process until smooth.

Add parsley and basil a little at a time, running the blender to combine the ingredients.

Add Parmesan cheese.

Add a pinch or two of salt to taste.

Try this hearty sauce on pasta (it serves four). You can also spread it on toasted French bread for *bruschetta*, on roasted vegetables or poultry. Add spoonfuls of pesto to flavor soups and sauces. Pesto can be frozen in small containers.

## GARDEN REFLECTION

### *Attention and Process*

GARDENS SURPRISE US with unexpected lessons, sometimes in unexpected places. I noticed a remarkable change when I came back to campus this fall. Next to our parking lot, the vacant lot on Franklin Street had become a garden. Seven-foot-tall sunflowers raised their heads along a new chain-link fence. Rows of corn, beans, greens, herbs, and flowers grew neatly inside. Someone—he looked like my colleague Doug—was standing in the midst of this paradise with a shovel and wheelbarrow. As I drove home trying to beat the traffic, I wondered if I'd been seeing things.

The next afternoon, before getting into my car, I walked over to the garden. The gate was swung open, and Doug waved me inside. He filled a plastic bag with red lettuce, kale, corn, turnips, yellow squash, and cucumbers as we walked among the rows of vegetables and flowers. Doug told me he'd gotten divorced that spring and moved into a small apartment above the corner coffee shop. Looking out his window at the vacant lot, he'd gotten an idea. "I just needed to dig in the dirt," he said, "to watch things grow." He found out the university owned the lot, so he'd asked people on campus, who finally said he could plant some vegetables. Over the summer Doug had rototilled the lot, brought in loads of fresh topsoil and steer manure. Some colleagues had given him seeds and plants, and by now everything was growing profusely.

"How do you get the plants to grow so well?" I asked.

"I give them lots of attention," he said. "Good soil, a couple applications of Miracle-Gro," he added. But mainly he spent every day cultivating his garden, breaking up the hard, barren soil and tending his plants.

While working in the garden, he had broken through his depression, watching the sun's warmth bring new seeds to life.

"This garden has been the best therapy," he explained. Out in the sun and fresh air, touching the earth, watching things grow put everything into perspective. After the faculty meeting that day, Doug had gone back to the garden, getting "grounded" again in the most essential way.

Now that his garden is flourishing, he's glad to share the harvest with colleagues. "Come back any time and help yourself," he said. "It's not the vegetables I wanted so much as the *process*. But now they're here, there's enough for all of us."

Offering to bring in some bags of manure from our horse, I went home with a bag of organic vegetables. As I prepared that night's salad, I thought about the abandoned lot that had blossomed into a garden. Two things Doug had said echoed in my mind: "I give them lots of attention." An important reminder: Giving our attention to something, really caring, produces remarkable results. Then there's process. In a culture that values products, material commodities, we often overlook the intangibles. The transforming power of process is one of life's best-kept secrets; commitment to process makes us stronger and wiser. For what we cultivate around us we also cultivate within us. Working in that vacant lot, Doug was growing a lot more than vegetables: he was cultivating his faith in life.

# Eleven

## November:
## Seasonal Feasts and
## Creative Interludes

To every thing there is a season,
And a time to every purpose under the heaven.

Ecclesiastes 3:1,
*King James Bible*
(1611)

OVEMBER IN MY garden brings cooler days and a blaze of
fall color. Early this month the Japanese maple in the south
yard turned a bright crimson. The crape myrtle in the inner
courtyard looked like a flame, its lower branches turned to gold and top
leaves orange and vermilion. In front the liquidambar is dropping star-
shaped red and gold leaves on the ground. Heart-shaped poplar leaves,
with red centers and rims of gold, fall like tiny Valentines among the sweet
alyssum.

Fall's transformations are remarkable. Down the street trees that were
green only a few days ago have turned to scarlet, burgundy, and plum. In
this autumnal symphony different trees produce different colors. The
leaves of the Lombardy poplar (*Populus nigra* Italica) are a deep wine red,
while one Japanese maple (*Acer palmatum*) is bright crimson. The diamond-
shaped leaves of the white birches (*Betula pendula*) have turned to gold,

ranging from butter yellow to hollandaise, mustard, and butterscotch, while the liquidambar leaves (*Liquidambar styraciflua*) span the spectrum from yellow to red. The trees' colorful performances build to their own crescendos. The crape myrtle first turned to gold, then became topped with red, until near the end of the month the entire tree was bright crimson. The next day, its performance concluded, the leaves began to fall.

Autumn comes at different times to different parts of the country. While colorful displays transform the New England countryside from September to October, on the West Coast most of our color comes in November. This month in the California wine country entire hillsides turn red and gold, rivaling New England in their autumn glory. Our friends Don and Barbara call these vineyards "more beautiful than Tuscany." No wonder Napa, Sonoma, and Mendocino are home to many artists' colonies.

Much of nature's artistry can be explained by science. Leaves of deciduous trees (from the Latin word meaning "to fall off") change color as part of their natural aging process. During the spring and summer the chlorophyll in these plants' leaves absorbs sunlight, using it to transform carbon dioxide and water to sugar and other carbohydrates. Plants use this process, known as photosynthesis, to produce food. When the days grow shorter and cooler in autumn, photosynthesis slows down. The leaves stop making chlorophyll and their green color disappears, revealing the carotenoids that have always been present but were masked by the bright green chlorophyll.

In some species—such as aspen, birch, hickory, black oak, American elm, beech, and willow—yellow colors predominate. In liquidambar, dogwood, Japanese maple, silver maple, mountain maple, and crape myrtle, leaves turn from yellow to red. The red colors are often caused by the synthesis of another kind of pigment, anthocyanins. Formation of these pigments is stimulated by bright sunlight and cool air temperatures. The most spectacular autumnal color develops in years when the weather is bright and cool (but not too cold, for freezing weather stops the process, causing the leaves to die and fall from the tree).[1]

The microclimates in your yard can produce different color intensities. The Japanese maple on the sunny, south side of our yard has turned bright red, while the other, in our shaded courtyard, is still green. Its top leaves are turning red, along with those of the nearby crape myrtle, because the highest branches are exposed to more daytime sun and cold night air.

## Seasonal Feasts and Holidays

THE END OF November for most Americans means Thanksgiving: roast turkey, dressing, cranberry sauce, yams, and pumpkin pie. With our multicultural population, Thanksgiving is one of the few rituals we share. I've asked my Italian, Polish, Latino, Chinese, and African American friends what they have for Thanksgiving dinner and, unless they are vegetarians, we all have virtually the same meal.

This year my parents flew up from San Clemente, my brother drove down from San Francisco, and we had Thanksgiving dinner at our house, with Bob cooking the turkey, stuffed with fresh pineapple sage and sprigs of rosemary from the garden. My parents made the dressing, my brother mashed the potatoes, and I baked the pies.

All the seasonal roasting and baking warms our homes and fills the air with the fragrance of herbs and spices. One simple baked dish I enjoy this time of year is stuffed acorn squash. Satisfying and nutritious, it bakes for an hour with no need for attention, freeing us to go about our holiday activities.

Gathering for November holiday feasts dates back long before the Pilgrims' first Thanksgiving. In medieval and Renaissance England, people began the month with Hallowmas or All Saints' Day on November 1.

### Baked Acorn Squash

*With a sharp knife, split the acorn squash horizontally into two equal portions. Scoop out the seeds and fibers from the center with a spoon. Place the squash, hollow sides up, in a baking dish sprayed with cooking oil and filled with 1 inch of water.*

*Fill the hollows with one small apple, sliced, along with raisins and walnuts or dried cranberries and pecans.*

*Sprinkle with sugar or cinnamon if desired. Cover the squash with a layer of foil and bake at 375 degrees for 1 hour. Serves two.*

*Tis not the food, but the*
*    content*
*That makes the Tables*
*    merriment.*
      ROBERT HERRICK,
    "Content, not cates,"
        ll. 1–2 (1648)

After 1605 a new holiday, Guy Fawkes Day, was pro-
claimed by the House of Commons to celebrate their
deliverance from the Gunpowder Plot. In late October
1605, Guy Fawkes and a group of conspirators con-
cealed thirty-six barrels of gunpowder beneath the
House of Lords, planning to blow up both houses of
Parliament on November 5, when King James I
presided at opening ceremonies. Since then November 5 has been celebrated with festive cakes, torchlight
processions, fireworks, and bonfires. The "guy" is
burnt in effigy, and children go from house to house
begging for pennies.[2]

November 11 was the final harvest feast of Martinmas, held at the same
time as an earlier Roman vintage feast. As their continental neighbors
were tasting their first vintage of the year, the English were storing meat
away for winter and celebrating the feast of St. Martin with drinking,
games, and a hearty meal of roast goose. At the end of the month came St.
Cecilia's Day on November 22, St. Clement's Day,
November 23; St. Catherine's Day, November 25; and
Tander or St. Andrew's Day, on November 30, which
began another round of drinking, feasting, games, and
celebration.[3]

Seasonal feasts gave each month purpose, taking
medieval men and women through the cycle of nature
and the liturgical year. In their world, as in our own,
holidays broke the routine of mundane tasks, reminding people of their part in a larger pattern of meaning.
Holiday feasts also relieved the monotony of medieval
and early Renaissance diets, which relied primarily
upon beans, peas, and bread, for only aristocrats ate
meat regularly.

At daybreak, most men and women of the time ate a
light meal of bread and ale. Then they had a more sub-
stantial meal of pottage (legume and vegetable stew),

**Feasting with the
King**

*Richard II, who ruled*
*England from 1377 to*
*1399, kept the most lavish*
*court of any English*
*monarch before or since.*
*His feasts were legendary.*
*Serving meals to 10,000*
*guests, he employed 300*
*cooks, who designed*
*special pies built like*

bread, and ale at 9:00 A.M., followed by a supper of pottage, bread, and ale at 4:00 P.M. The next day they ate more bread, ale, and pottage. For flavor they seasoned their pottage with thyme, basil, mint, and marigolds.

In the warmer months they enjoyed fruit in season, stewing their apples, berries, and pears in heavy syrup or baking them in pies. They also drank apple cider and pear cider, or perry, and the upper classes drank wine. But in the winter, except for feasts, most people's diets revolved around pottage, bread, and ale.[4]

Medieval feasts were an extravagant diversion, with elaborate dishes to delight the eye as well as the palate. During the time of Richard II, feasts lasted all day. In the Renaissance as well, seasonal feasts offered a break from dull routine, bringing warmth and color to the dark days of winter.

*castles and roasted great quantities of meat. One feast in 1387 for Richard and his uncle, the duke of Lancaster, included 16 oxen, 120 sheep, 12 boars, 14 calves, 140 pigs, 50 swans, 210 geese, 58 capons, 60 dozen hens, 400 rabbits, 11,000 eggs, and a variety of fowl, as well as enormous quantities of fruits and vegetables. Feasts began about 10:30 in the morning and lasted all day. They included three complete meals, each with eight to twelve courses. After the last course was served at sunset, the king bid his satiated guests good night.[5]*

## GARDEN GROWTH

AS HOLIDAY FEASTING begins, garden growth slows in November, even in Northern California. I still have rosemary, thyme, and sage in my kitchen garden, along with some fresh greens. Four of the garlic cloves I planted last month have sent up shoots. Early this month, when we had our first frost warning, I cut the last clusters of green tomatoes from the vines on our south garden wall, bringing them inside to ripen on the windowsills.

My chrysanthemums are still blooming in shades of bronze, lilac, rose, and gold, as is the purple velvet salvia. Echoing their autumnal colors, the ajuga nearby has produced dozens of tiny purple spires.

In the friendship garden the blooms on Rhonda's society garlic have finally surrendered to time, leaving only a scattering of autumn leaves and

*These fresh beauties (we
can prove)
Once were Virgins sick of
love,
Turn'd to Flowers. Still in
some
Colours goe, and colours
come.*

ROBERT HERRICK,
"Why Flowers change
colour"
ll. 1–4 (1648)

foliage from early bulbs. In the backyard the hydrangea blossoms have turned to deep rose and rust.

My November garden offers a rare convergence of seasons. Acceding to autumn, the rose of Sharon's leaves have yellowed and fallen away, leaving only bare branches. The Brazilian plume flower on its right looks back to summer, filled with bright pink blossoms, while the heavenly bamboo on the other side is decked with bronze foliage and clusters of winter red berries.

Autumn leaves cover the ground in shades of rust and gold, falling faster than I can keep up with them. One day I will rake up three bushel baskets in the front yard. The next morning that many more leaves will have fallen. But even though each day's efforts are effaced by the wind, no time in the garden is ever wasted. The autumn leaves will make valuable compost, and I find working out in the crisp November air invigorating. It clears my mind, renews my spirit, and provides light exercise to balance all the holiday feasting.

## GARDENING AS SPIRITUAL PRACTICE

### Mindfulness in Small Things

WORKING IN OUR gardens sharpens our powers of observation, making us mindful of small things. If the leaves are still turning in your area, let them inspire your contemplative practice this month. Notice the subtle differences in color and intensity. Look for the different effects of microclimates in your yard. Do you see something you hadn't noticed before?

November weather brings powerful contrasts. It rained all day yesterday and last night. Today the sun is breaking through the clouds. As I look out my study window, a tiny hummingbird darts among the spires of purple salvia. The last golden leaves on the liquidambar sparkle in the morning sun. Raindrops shimmer on the bronze leaves of the heavenly bamboo. I pause in a moment of silent appreciation.

Contrast seasons our admiration and awakens our senses. Even winter weather brings its blessings. If every day were filled with sunshine, we would not appreciate the simple beauty of the morning after a winter storm.

### *Personal Exercise: Cultivating Mindfulness*

THIS TIME OF year we're often so busy, caught up in holiday tasks, that we miss the small moments of beauty around us. Today give yourself the gift of mindfulness. If the weather permits, take a walk in your garden. If not, take a long look out your window at the November landscape.

*Thus I, easy philosopher,*
*Among the birds and trees*
*confer,*
*And little now to make*
*me wants,*
*Or of the fowls or of the*
*plants:*
*Give me but wings as*
*they, and I*
*Straight floating on the*
*air shall fly;*
*Or turn me but, and you*
*shall see*
*I was but an inverted tree.*
ANDREW MARVELL,
"Upon Appleton House,"
ll. 561–68 (c. 1652)

- What subtle changes do you see?
- What details awaken your appreciation?
- What lessons does the landscape hold for you?

### *Following Your Creative Practice*

IN NOVEMBER THE days grow noticeably shorter and winter weather begins in earnest. When rain or snowstorms keep us from our gardens, we can enjoy the creative practice we began last month.

Any creative practice brings its lessons in personal growth. To discover more about your own lessons, ask yourself the following questions:

- What practice did I choose: painting, music, woodworking, needlework, weaving, something else?

🌺 What do I enjoy *most* about my creative practice and why?

🌺 What do I enjoy *least* about this practice and why?

🌺 What have I learned from the practice? About myself? About the patterns around me?

A favorite creative practice for me is needlepoint. I like to make tapestries in Renaissance bargello or medieval herbs and flowers. I love the relaxing, repetitive motions, the reverie of working with colorful yarns, and watching the slow emergence of patterns before my eyes. The process is at least as important as the final result, offering me a quiet respite from the day's activities and the tangible reassurance that my efforts combine to create new patterns of beauty, if only in the small canvas I hold before me.

So much of what I do in my life as a college professor is intangible and ephemeral: lectures, meetings, calls and e-mails, papers to grade and return. Meanwhile, my tapestry slowly evolves into colorful patterns of strawberries, daisies, daffodils, and fleur-de-lis, reminding me of the progression of flowers and harvests in the garden of my life: the patterns blossoming in my yard and the more subtle patterns that blossom within and around me in love, friendship, and personal growth.

Whatever your creative practice, it can become for you an ongoing spiritual exercise, making you more aware of the meaningful patterns in your life while helping you develop greater faith in the creative process.

## Contemplation and Creativity

THE WINTER SEASON itself offers vital lessons in creativity that are too often overlooked. Spring and summer growth and autumn harvests are obvious. Winter's lessons are powerful but more subtle. By now most of us are composting spent vegetables and annuals, preparing our gardens for winter. Yet as one garden year ends, another begins with this season of silence and contemplation. Even after we add organic matter to enrich the soil and put away our garden tools, we participate in a creative cycle that transcends us, includes us, is beyond our control.

The better we understand nature's creative process, the better we can cooperate with it to improve our gardens and our lives. In our gardens there is a time for every purpose: a time to plant and a time to reap, a time to work and a time to wait. Composting our autumn leaves, spreading manure on our garden beds, or even planting our last spring bulbs, we must leave them alone for a while and let nature take its course. This is the lesson of winter.

## Allowing Ourselves the Grace of Contemplation

AS WE SET most of our active garden work aside, very little seems to be happening. Yet valuable growth can emerge from winter dormancy, in both our gardens and our lives. Contemplation is an essential part of any creative process, a fact known by centuries of artists, scientists, writers, and composers whose inspirations occurred on walks, in dreams, or in moments of reverie.

From the Middle Ages to today artists in all fields have followed the fourfold creative process of

- *preparation*, a period of active work
- *incubation*, a period of rest or contemplation
- *inspiration*, the creative insight
- *verification*, following up on the inspiration with renewed action.

*Lay by the good a while;*
*  a resting field*
*Will, after ease, a richer*
*  harvest yield.*
          ROBERT HERRICK,
          "Rest Refreshes,"
          ll. 1–2 (1648)

Hard work and the best of intentions are not enough. The creative process *requires* a period of repose. Without incubation there can be no inspiration.

## Personal Exercise: Working with the Cycles

WE, TOO, CAN follow the creative process, combining our efforts with time and nature's cycles to produce new insights, inspirations, and solutions to current problems. We have only to conquer our impatience, release the compulsion to control, make room in our lives for contemplation, and trust the process.

You can begin by considering the following questions:

☙ Can you identify one creative cycle in your garden? How did action alternate with incubation to create something new?

☙ Can you identify one current creative cycle in your life:
- in your work?
- in your relationship?
- in your family life?
- in your exercise program?
- in a current challenge?
- or in some other area?

☙ What can you do to combine action with incubation in this area? How can you detach from active involvement long enough to gain greater insight and perspective?

- Could you take a break when you reach an impasse with a project, coming back to it when your mind is fresh?
- Could your family or relationship benefit from some unstructured time away from daily routine: an afternoon outing or brief vacation?
- Could your exercise program benefit from peaceful walking, yoga, or tai chi between aerobic training days?
- Could you give yourself more margins: quiet periods of rest and reflection to cultivate your faith?

*Take of this grain, which*
*in my garden grows,*
*And grows for you;*
*Make bread of it: and*
*that repose*
*And peace, which ev'ry*
*where*
*With so much*
*earnestnesse you do*
*pursue,*
*Is onely there.*
GEORGE HERBERT,
"Peace,"
ll. 37–42 (1633)

This lesson offers us a gentle reminder as we enter the winter holiday season, with its festive and often frantic celebration. Our culture is always so "busy" that it takes courage to allow ourselves the grace of contemplation. But by combining periods of action and contemplation, you can actually *save* time: opening yourself up to inspiration, creative breakthroughs, and new possibilities while cultivating greater peace in your life.

## GARDEN TASKS

### *Planting Cool-Weather Crops and Honoring the Season*

NOVEMBER GREETS GARDENERS across the country at different points of nature's cycle. If you live in zones 1 through 6, in New England, the Midwest, or the Rocky Mountain states, you are probably mulching your garden for winter. In zones 7 through 11, you can still plant cool-weather vegetables: fava beans, broccoli, cabbage, chard, spinach, lettuce, and other greens, although in zones 7 and 8, and perhaps 9, you'll need to use a cold frame. This month I moved a planter from the courtyard to the warmer south side of the house and sowed some mesclun green seeds there.

As long as the ground isn't frozen in your region, you can sow wild-flower seeds. Cultivate the soil and cover the seeds lightly with earth. These buried treasures will emerge when the world comes to life again next spring.

Local guides for my zone 9 garden say I can also set out cool-weather bedding plants: pansies, snapdragons, English primroses, and stock. I love the velvet faces of pansies, and the handful I planted this fall have barely begun to bloom. Last week the garden center had racks of them in color-ful six-packs. On campus this week gardeners were setting out over three hundred pansies in rows along the sidewalk between my classroom and the mission church. But for me planting new flowers seems suddenly out of season.

I remember the hard frosts of last December that blasted so many ferns and shrubs, brought the salvia to the ground, and turned our birdbaths to solid ice. But more than that, I appreciate the winter cycle, with its time for contemplation and quiet celebration, a season to reflect and prepare for another cycle of growth.

The drive for continuous productivity is one of the banes of our time.

People around me call rare moments of repose downtime, a word used when computers break down. But the metaphor is inaccurate, for we are not machines. Before the industrial and information ages, men and women were more in touch with nature's cycles. Working the soil for their livelihood, they could see how each action, each season, leads to the next in spirals of growth, sometimes obvious, sometimes—like the bulbs I planted last month—occurring secretly, underground and unobserved.

November for me marks a time to turn within, to prepare the soil for next year's garden and do some long-range planning. As much as I love pansies, I want them next spring, not now. Delayed gratification gives us time to reflect, time to dream, to return to our roots like perennials so that we, too, may emerge in new seasons of growth. The lives of plants, poetry, and music are not linear but cyclical, not continuous commotion but alternations of sound and silence that follow nature's rhythms and bring greater harmony to our lives.

## Tending and Protecting Your Plants

AS YOUR GARDEN slows down this month, remember to fertilize your remaining flowers and vegetables. If it doesn't rain much in your region, you'll want to keep watering, but in most areas the winter weather prompts us to cut back or turn off our automatic systems altogether. This is the time many of us put our gardens to bed, clearing out debris; cleaning and stowing away stakes, trellises, and clay pots; putting spent annuals into the compost bin.

When my chrysanthemums and other herbaceous perennials stop blooming, I'll cut them back to about six inches from the ground, composting these clippings as well. If you haven't already done so, stop deadheading your roses to let them finish blooming and prepare for winter. Some varieties have bright orange rose hips that add to the seasonal display. In colder regions it's time to protect your roses and other vulnerable plants with winter coverings and mulch.

## Protecting Plants from Frost and Cold Weather

MANY OF US will need to begin protecting our plants from frost, which occurs on cold, clear nights with no cloud cover. Here are some guidelines:

- *Protect container plants* by moving them inside or next to the house. If you leave them out, wrap the containers in insulation material, newspaper, or burlap.
- *Cover somewhat hardy plants with evergreen boughs.* Place boughs over and around the plants you want to protect. You may need to anchor the boughs' ends in the ground and tie them in place.
- *Build a frost protection barrier around a plant* with a frame of chicken wire anchored to the ground and covered with a layer of insulation material, burlap, or newspaper. Make sure your covering allows air circulation.
- *Guard citrus trees and other vulnerable plants from frost at night* with a layer of insulation, plastic canvas, or even newspaper. Use stakes to keep the covering from touching the plant, so it won't freeze. Remove the covering in the morning.
- *Mulch around broadleaf evergreens and other shrubs.* Spread a six- to twelve-inch layer of mulch around the plants to insulate their roots, protecting them from alternating freezes and thaws.
- *Protect roses from the cold* by making a mound of soil from six to twelve inches high around their stems. If your region gets especially cold, add a six- to twelve-inch layer of mulch over the mound.

## Cultivating the Soil

NOVEMBER BRINGS RAIN showers to my Northern California garden, but when weather allows this is a good time to prepare our garden beds for the next growing season. At a recent class at the Gamble Garden Center in Palo Alto, the horticulturist Don Ellis walked over to one of the beds, scooped up a handful of soil, and held it out. It was as crumbly as pie dough, remarkable because the natural state of the soil around here is heavy clay. Inspired to amend my soil as well, I dug the north garden beds, returning the spent bean plants to the soil as green manure, then worked in a bag of organic compost.

When we add compost we also need to add nitrogen because most compost uses nitrogen to decompose, removing this nutrient from the soil. To find out how much nitrogen to add, I bought a soil test kit at the garden supply store. Feeling like a scientist, I removed the four plastic tubes from the kit, added the right amount of water and soil from the garden beds, and awaited the results. The pH level was fine. The mixture of soil, water, and chemicals in the tube bubbled up and turned a pale green for 7.0 or neutral pH, midway between acid and alkaline. But the other tests revealed very low nitrogen and potassium. I blinked in amazement. The soil was anemic! No wonder the beans had never thrived there.

*But sweet Vicissitudes of Rest and Toyl Make easy Labour, and renew the Soil.*

John Dryden,
*Virgil's Georgics,*
I, ll. 116–17 (1697)

Gardening puts us in touch with the system of largely invisible biochemical interactions that supports all life on this planet. Plants need more than soil, sunshine, and water. Their soil—like the foods we eat—must provide them with essential nutrients. That means seventeen essential elements: the nine macronutrients—nitrogen, phosphorus, potassium, sulfur, calcium, magnesium, oxygen, carbon, and hydrogen—required in large amounts; and the eight micronutrients or trace minerals—iron, copper, zinc, nickel, boron, manganese, chlorine, and molybdenum—required in smaller amounts.[6]

## Testing and Treating the Soil

AS YOU'RE DIGGING and composting your garden beds for winter, it's a good time to test your soil. You can buy an inexpensive test kit at a hardware store or garden center. Get one that tests not only pH (acidity-alkalinity) but also the three essential nutrients: nitrogen, phosphorus, and potassium (N, P, and K). Most test kits will have one or more containers and chemical capsules along with instructions.

- *pH*. Soils range from alkaline (8.0) to very acidic (5.0), with 7.0 as neutral. Some plants prefer acidic soil, others prefer alkaline, but most prefer something in between. The test kit should list common plants and their pH preferences, which you can compare with your soil's pH level. To raise your soil's acid level, add calcium sulfate, iron sulfate, or acidic organic matter such as bark or manure. To make your soil more alkaline, add bonemeal, wood ash, or dolomite limestone. Check with your local garden center for recommended amounts.
- *Nitrogen*. If your nitrogen level is low, add sulfate of ammonia or aged manure to your soil.
- *Phosphorus*. If your phosphorus level is low, you can add superphosphate or organic sources such as bone, hoof, or horn meal.
- *Potassium*. If your potassium level is low, you can add potassium sulfate, wood ash, or leaves and leaf mold.[7]

Like much of the soil in California, my garden beds were low in the most essential element, nitrogen, which is part of an ongoing cycle. When organic matter decays it is broken down by bacteria in the soil into ammonium, which is then oxidized into nitrites and converted by other bacteria into nitrates, a form of nitrogen absorbed by the roots of plants.[8] Our horse, Rusty, loves to eat grass, which has taken in nitrogen from the soil. She digests her food and excretes manure onto the ground, where bacteria in the soil break it down into ammonium, which is converted by soil bacteria into nitrites, then into nitrates, which support the growth of new grass. Rusty eats the grass, and the cycle begins again.

Nitrogen is also added to the soil by "green manure," plants such as alfalfa (*Medicago sativa*), clover (*Trifolium*), peas (*Pisum*), and beans (*Phaseolus*) that develop symbiotic relationships with nitrogen-fixing bacteria (*Rhizobium* and *Bradyrhizobium*). For centuries wise farmers have rotated their crops, alternating grain with these helpful plants to enrich their soil.[9]

## Nitrogen-Carbon Interaction in Your Garden

NITROGEN HELPS ORGANIC matter decompose, combining with carbon in the decomposition process. The higher the degree of carbon, the greater the microbial activity, and the greater the amount of nitrogen used. In our gardens, as in our lives, balance is essential. If the organic matter you add to your soil contains lots of carbon, it uses more nitrogen than it contributes, resulting in soil depletion.

When you add compost to your garden, you'll want to choose an organic source with a carbon-nitrogen ratio lower than 20 to 1. Cow, pig, sheep, and chicken manure are all usually 20 to 1 or lower, whereas horse manure is 50 to 1. Sawdust or wood chips can be as high as 400 to 1. If you use these you'll need to supply additional nitrogen to avoid a soil deficiency.[10]

I called the master gardener hotline at our local University of California extension, where a friendly master gardener told me to add sulfate of ammonia, according to label recommendations, to the garden beds now, dig in some manure, and let the soil bacteria begin breaking it down into usable nitrogen. To raise the potassium level he said to "work leaves into the soil now and let them break down." The autumn leaves in our yard are a good source of potassium.[11]

Last Sunday afternoon I put on my oldest work clothes and drove out to Rusty's barn. Wearing a cowboy hat, sunglasses, and a blue bandanna over my nose, I shoveled horse manure into five plastic garbage bags. Bob laughed and said I looked like a bandit. But the point was to collect the manure without inhaling it. Loading the bags into a weathered red wheelbarrow, I put them into the trunk of my car and took off my bandanna for the drive home. I unloaded the bags, spread two on each garden bed, and dumped the last one into the compost bin. The forecast for the next few days predicted rain, which has already begun to break down the manure to release its nutrients into the soil.

Gardeners have been enriching the soil with manure since the Middle Ages, when people dug trenches around their gardens and filled them with dung from farm animals, household rubbish, and even "night soil" from chamber pots. In his 1580 gardening guide, Thomas Tusser recommended this practice for November:

> If garden require it, now trench it ye may,
> One trench not a yard from another go lay;
> Which being well filled with muck by and by,
> Go cover with mould for a season to ly. ·
> . . .
> Foule privies are now to be clensed . . .
> Which buried in garden, in trenches alowe
> Shall make very many things better to growe.

Two centuries later in America, Thomas Jefferson carefully cultivated the soil in his garden at Monticello. We still have a letter he wrote to his daughter Martha in July 1793, while handling the new nation's business in Philadelphia: "We will try this winter to cover our garden with a heavy coating of manure. When the earth is rich, it bids defiance to droughts, yields in abundance, and of the best quality."[12]

For thousands of years gardeners have returned waste products to the soil in this ongoing cycle of renewal. Virgil's *Georgics*, written around 30 B.C.E., tells gardeners to add manure and ashes to their soil. One medieval manuscript shows gardeners in Provence pruning grapevines, then burning them in bonfires and enriching the soil with the ashes.[13] Organic gardeners still use wood ash to add potassium sulfate to the soil, along with bonemeal, which adds phosphorus.

> *Yet sprinkle sordid Ashes*
> *all around,*
> *And load with fat'ning*
> *Dung thy fallow*
> *Ground.*
> JOHN DRYDEN,
> *Virgil's Georgics,*
> II, ll. 118–19 (1697)

### Harvesting Late Vegetables and Autumn Leaves

IN AREAS WITH mild winters you may still be harvesting some late vegetables. In the planter on the south wall, I grew some organic spinach, which was wonderful in salads and stir-fry dishes. I also brought the remaining green tomatoes inside to ripen. For the rest of the month the harvest from my kitchen garden has been limited to green onions, sage leaves, and sprigs of rosemary.

The main harvest in my yard now is leaves, bushel baskets full, from the liquidambar, maple, birch, poplar, and crape myrtle trees as well as many deciduous shrubs in our yard. I followed the master gardener's advice and dug some liquidambar leaves into the north garden beds. Then I took the five plastic bags I'd used to haul Rusty's manure and filled them with leaves, poked a few holes for ventilation, and set them aside to make leaf mold.[14]

Working with nature's cycles, we can use time as an essential ingredient to amend the soil. If we do some digging and cultivating now, the winter frost will break down the soil and improve its texture. If we spread a layer of manure over the soil, the winter weather will help it break down and release nutrients into the soil. Leaf mold started now will be ready to nourish our summer gardens. In all these cases we just start the process and let nature do the rest.

## GARDEN REFLECTION

### The Lesson of November

As NOVEMBER USHERS in the season of winter, we can take time to find our own meaning in this contemplative season. For, as our gardens remind us, every season has its purpose. As Dryden said in his translation of Virgil:

> Thus ev'ry sev'ral Season is employ'd:
> Some spent in Toyl, and some in Ease enjoy'd.
> The yeaning Ewes prevent the springing Year;
> The laded Boughs their fruits in Autumn bear.
> 'Tis then the Vine her liquid Harvest yields,
> Bak'd in the Sun-shine of ascending Fields.

The Winter comes, and then the falling Mast,
For greedy Swine, provides a full repast.
Then Olives, ground in Mills, their fatness boast;
And winter Fruits are mellow'd by the Frost.[15]

Like the seasons of the year, each season of our lives has its meaning and purpose.

Evolving through the seasons, each of us creates his or her life pattern. C. S. Lewis wrote that "humanity does not pass through phases as a train passes through stations: being alive, it has the privilege of always moving yet never leaving anything behind. Whatever we have been, in some sort we are still."[16]

With its fallen leaves, nitrogen cycles, and keen evidence of the changing seasons, November reminds me of the interconnectedness of life: "Whatever we have been, in some sort we are still." The seasons of our garden year affirm the underlying process of creative growth in which we cultivate our gardens and our lives. November, the advent of winter, is an essential part of the pattern, a time to return to our roots like so many perennials, to listen to our hearts, seek moments of beauty, and follow the inner light.

# WINTER

# Winter Garden Checklist

## EARLY WINTER

**Plant:** Before the ground freezes, plant bare-root roses, trees, and shrubs, and the last spring bulbs. Order seeds for spring planting. If your winters are mild, set out cool-weather bedding plants.

**Water:** Continue to water as needed. Deep-water young trees and bushes before the ground freezes.

**Weed:** Clean up weeds and plant debris.

**Feed:** Feed any fall-planted annuals and vegetables. Do not fertilize roses until spring. Fertilize evergreens, and feed deciduous trees when dormant.

**Cut back:** Chrysanthemums and other herbaceous perennials when they have finished blooming as well as any spent plants in the vegetable garden. Clean and oil garden tools. Sharpen pruning tools. Prune berry bushes, grapevines, and fruit trees when they are dormant. Use evergreen and holly prunings to decorate for the holidays.

**Cultivate:** Rake up and compost fallen leaves. Dig and cultivate empty garden areas. Test your soil and amend as needed. Add organic matter: manure and compost. Let the manure settle in for one or two months before planting. In colder regions protect roses and other vulnerable plants with winter covering, mulch around trees and shrubs, and stake young trees. Check coverings periodically. Knock snow off small trees and shrubs to keep the weight from breaking their branches. Start planning next year's garden.

**Harvest:** In milder areas harvest late fruits and vegetables. Check winter squash and root crops in storage.

**Watch for:** In milder regions pick up fallen fruit and leaves. Check and treat for snails, slugs, and mildew. If you have brought in dahlia tubers and other tender bulbs for the winter, check them for rot or mildew. Also check stored winter squash and root crops. Discard any affected bulbs or vegetables.

**Enjoy:** Enjoy late harvests and holidays; looking back on this year's garden and planning for the next.

## MIDWINTER

❧ *Plant:* If the ground is not frozen, plant bare-root roses, hedges, trees, fruit trees, and shrubs. Order seeds for spring planting. In colder areas sow hardy annuals and early vegetables in indoor flats. In milder areas sow hardy annuals, early peas, cabbage, lettuce, spinach, radishes, and onion sets outdoors.

❧ *Water:* Continue to water as needed.

❧ *Weed:* Clean up weeds and plant debris.

❧ *Feed:* Feed emerging bulbs with bonemeal, and fertilize any new vegetables and hardy annuals. Feed fruit trees around the roots with a high-potash fertilizer when they are still dormant and azaleas, camellias, and rhododendrons with acid food to support spring growth. Feed established trees and shrubs when they begin to grow.

❧ *Cut back:* Clean and oil garden tools. Sharpen pruning tools. Prune fruit trees and rosebushes while they are dormant. Do not prune climbers and ramblers until late summer.

❧ *Cultivate:* Dig and cultivate the soil for spring planting. Test your soil and amend as needed. Add organic matter: manure and compost. Let the manure settle in for one or two months before planting. In colder regions check protective coverings for roses and other vulnerable plants and any staked trees and shrubs. Knock snow off small trees and shrubs to keep the weight from breaking their branches. Review your garden's design and last year's gardening successes and failures as you plan for the year ahead.

❧ *Harvest:* Harvest stored winter squash, root crops, and thinnings from early greens.

❧ *Watch for:* If you have brought in dahlia tubers and other tender bulbs for the winter, check them for rot or mildew. Also check stored winter squash and root crops. Discard any affected bulbs or vegetables. Pick up fallen leaves around the garden. If snails and slugs are attacking emerging bulbs and other plants, dust the tops with wood ashes or use snail bait.

❧ *Enjoy:* Enjoy celebrating new beginnings and laying out your plans for the garden year ahead.

## L A T E   W I N T E R

✁ *Plant:* If the ground is not frozen, plant bare-root roses and trees, and container-grown trees and shrubs. Order roses, seeds, and bedding plants for spring planting. In colder areas sow hardy annuals and early vegetables indoors or in your garden with cloches or cold frames. In milder areas sow sweet peas and hardy annuals, early peas, cabbage, lettuce, spinach, radishes, and onion sets outdoors. When weather permits plant spring perennials.

✁ *Water:* Continue to water as needed. In milder regions water new plantings and shallow-rooted azaleas.

✁ *Weed:* Pull weeds while their roots are still small.

✁ *Feed:* Feed emerging bulbs with bonemeal. Feed roses, trees, and shrubs as new growth begins and fruit trees before they bloom. Fertilize hardy annuals, newly planted vegetables, and new plantings in beds and borders. Wait to feed azaleas, camellias, and rhododendrons until after they bloom.

✁ *Cut back:* Prune rose bushes while they are dormant. Do not prune climbers and ramblers until late summer. Prune berry bushes, grapevines, and fruit trees as well as other dormant bushes and shrubs. Wait to prune spring-flowering shrubs until after they bloom. Deadhead spent blossoms from bulbs but leave foliage in place. Thin early vegetables.

✁ *Cultivate:* Dig and cultivate the soil, adding organic matter. In colder regions check protective coverings for roses and other vulnerable plants, and any staked trees and shrubs. Knock snow off small trees and shrubs to keep the weight from breaking their branches. When the weather warms up, begin removing protective coverings. In milder areas watch for late frosts and cover vulnerable plants.

✁ *Harvest:* Harvest stored winter squash, root crops, and thinnings from early greens.

✁ *Watch for:* If you have brought in dahlia tubers and other tender bulbs for the winter, check them for rot or mildew. Also check stored winter squash and root crops. Discard any affected bulbs or vegetables. Pick up fallen leaves. Treat snails and slugs with traps or bait.

✁ *Enjoy:* Enjoy the earliest signs of spring in your garden.

# TWELVE

## December: Celebration,
## Discipline, and Dreams

At Christmas I no more desire a rose
Than wish a snow in May's new-fangled shows,
But like of each thing that in season grows.

WILLIAM SHAKESPEARE,
*Love's Labour's Lost*,
I.1, ll. 105–7 (c. 1594)

GARDENING REVEALS THE lessons in every season. In mid-December much of my garden is dormant and the trees along our street are bare, their trunks and branches like gnarled hands reaching for the sky.

When I lived in Southern California, I barely noticed the winter. Christmas with my parents in San Clemente seems almost tropical. Azaleas and oleanders brighten the landscape, and children frolic on the beach. At home in the San Francisco Bay area, many trees and shrubs are bare and garden life slows down. Flocks of Canada geese fly south, their calls echoing overhead, and the colder, shorter days call us to go within, to cultivate our inner gardens.

The brightest colors on my street are holiday decorations: wreaths and colored lights set against the winter grays and browns. Centuries ago, in a world without electric lights, how much greater was the need to compensate, to add holiday color and warmth to this dark time of year, when, as Shakespeare wrote:

> . . . yellow leaves, or none, or few, do hang
> Upon those boughs which shake against the cold,
> Bare ruined choirs, where late the sweet birds sang.[1]

In the Middle Ages and the Renaissance, winter was a cold, dark season, when people stayed close to home, sustaining themselves on stored provisions and gathering around the fire. They treated their winter chills with herbal remedies and sprinkled dried herbs in their food and around the house, recalling the flavors and scents of lost summer days.

Holiday festivities brightened the darkness, blending Christian and agrarian rituals into a round of seasonal celebrations at risk today of being obscured by all the holiday shopping and spending. But beneath the glittering commercial wrappings, enduring traditions affirm the recurring miracle of life.

*In Genial Winter, Swains*
*enjoy their Store.*
*Forget their Hardships,*
*and recruit for more.*
*The Farmer to full Bowls*
*invites his Friends,*
*And what he got with*
*Pains, with Pleasure*
*spends.*

JOHN DRYDEN,
*Virgil's Georgics,*
I, ll. 403–6 (1697)

In mid-December, around the winter solstice, the Jewish people celebrate the eight days of Hanukkah, the festival of lights. Around the same time many ancient peoples throughout Northern Europe celebrated the solstice by lighting fires against the darkness. The Scandinavians celebrated the birth of the sun god, Woden, with evergreen trees and gifts, feasting on roast boar's head and burning the yule log. The ancient Romans celebrated the end of December as Saturnalia, the feast of Saturn, god of seeds, honoring the end of the agrarian cycle with drinking and feasting.[2]

Christian feasts became interwoven with these earlier traditions. December 21, the winter solstice, became St. Thomas Day, marked by the old rhyme:

St. Thomas grey, St. Thomas grey,
Longest night and shortest day.[3]

Yule logs, candles, gifts, and evergreen trees became associated with St. Nicholas in medieval Christmas celebrations. Men and women decorated their homes with holly and evergreen boughs, rosemary, box, mistletoe, ivy, and yew, ancient symbols of life that endured through the dead of winter.[4] They celebrated the four Sundays before Christmas as Advent, a time of spiritual preparation and anticipation. During my childhood in Germany I learned to make traditional Advent wreaths of evergreen boughs and four red candles, lighting one candle on the first Sunday of Advent, then adding another on each successive Sunday until all four would shine the Sunday before Christmas.

From the Middle Ages through the Renaissance, December was celebrated with a parade of holidays: St. Nicholas (December 6), St. Thomas (December 21), Christmas Eve (December 24), Christmas Day (December 25), St. Stephen's (December 26), Childermas or Holy Innocents (December 28), and December 31, the last day of the old year, followed by the Feast of the Circumcision on New Year's Day and Epiphany (Twelfth Night), marking the visit of the three wise men to Bethlehem, on January 6.

English holiday celebrations continued the traditional feasting of pagan antiquity. Henry VIII celebrated Christmas with hunting, feasting, dancing, and games. His gardens were decorated with silk flowers, gold, and pearls. Dressed in their best velvets, lords and ladies ate boar's head, capons, goose, swan, pheasant, spiced pies, plum pudding, and roast peacock. On December 26, St. Stephen's Day, the feasting continued with pies made from leftover Christmas goose. The Twelve Days of Christmas, between December 25 and Epiphany, January 6, were filled with parties, plays, masques, and revels at court and celebration throughout the countryside. New Year's

*At Christmas play and make good cheere, For Christmas comes but once a yeare.*
THOMAS TUSSER,
*Five Hundred Points of Good Husbandry* (1580)

Eve was a time for masked balls, feasting, songs, predictions, and drinking wassail around the fire until the church bells rang at midnight.[5]

Drinking healths or wassail was another English Christmas custom, from the Old English *wes hal*, which meant "may you be hale" (healthy). Even now, my colleague Richard Osberg, professor of medieval literature and his wife, Sally, host an annual Christmas party where they serve a traditional wassail punch.

Many early English men and women took their holiday celebrations into their gardens, toasting their fruit trees that they might bear abundantly in the coming year. Whether we wassail them or not, we are sure to remember our gardens as we acknowledge our many gifts this holiday season.

*And let the russet Swaines*
  *the Plough*
*and Harrow hand up*
  *resting now;*
*and to the Bag-pipe all*
  *addresse;*
*Till sleep takes place of*
  *wearinesse.*
*And thus, throughout,*
  *with Christmas playes*
*Frolic the full twelve*
  *Holy-dayes.*
    ROBERT HERRICK,
  "A New-yeares gift sent
  to Sir Simeon Steward,"
      ll. 45–50 (1648)

## GARDEN GROWTH

THIS MONTH THE palette in my garden has turned to reds, golds, and browns. The hydrangea blossoms have gone cherry red, their leaves Naples yellow. Crimson berries stud the heavenly bamboo. The nearby liquidambar, the last tree in my yard to change, is a glorious pale gold, the color of medieval parchment, against a gray December sky.

Garden growth in December is slow and subtle. Beneath the surface bulbs quietly prepare for spring, some sending up their first green foliage, others still asleep. Many perennials, too, are engrossed in their long winter's nap. Even those few plants still growing have slowed down. By December 17 the mesclun greens I'd sowed in a planter on the warm south side of the house had put up a few tiny green leaves. By the end of the month, the leaves were not much bigger. (I should probably put them in a cold frame.) Winter days, with their cooler weather and few hours of sunshine, do not promote vigorous growth.

Since mid-December we've had frosts and some high winds. Lots of leaves have been falling. I raked up three bushel baskets of leaves in the front yard, put them in plastic bags, and added them to the leaf mold pile. The next day I raked up four more. Hidden away in a corner of the yard near the toolshed are bags of leaf mold, where time quietly does its work, turning the leaves into potassium-rich compost to return to the soil in a few months. So much about gardening puts us in touch with the circle of life that includes all creation. Making compost and leaf mold is one such process, bringing nature's eternal cycle home.

*Wassaile the Trees, that*
*they may beare*
*You many a Plum, and*
*many a Peare:*
*For more or lesse fruits*
*they will bring,*
*As you doe give them*
*wassailing.*
ROBERT HERRICK,
"Another,"
ll. 1–4 (1648)

## GARDENING AS SPIRITUAL PRACTICE

### Christmas Cards and Friendship Gardens

THIS IS THE season to exchange holiday greetings with friends and family, a ritual that can be either compulsive or contemplative, depending upon how much we crowd into our schedules.

December haunts many of us with the ghosts of Christmas past. If our memories are happy, we long for the simple joys of childhood: the visits with our grandparents, the glow of bubble lights on the Christmas tree, gingerbread cookies, frolicking in the snow, and playing with our new puppy on Christmas morning. If our Christmases past were painful, we can be drawn back into unhappy patterns. Either way many people rush about during the holidays, searching the crowded shops and brightly colored packages for something they've left behind.

The wisdom of gardening makes us ever mindful of the present moment: finding beauty and meaning not in what we don't have but in what is right before our eyes: preparing the soil, tending our last remaining plants, putting our garden tools in order, enjoying the fragrant evergreen boughs on the mantle, or gazing at the quiet beauty of a December landscape.

The garden in winter reminds us that less is more. Our gardens and our lives are healthier when they're not overcrowded, and there is much to appreciate that can't immediately be seen. This year I've tried my best to simplify, choosing holiday rituals not out of habit or compulsion but because I find meaning in them.

*Who would have thought
my shrivel'd heart
Could have recover'd
greennesse? It was gone
Quite under ground; as
flowers depart
To see their mother-root,
when they have blown;
Where they together
All the hard weather,
Dead to the world, keep
house unknown.*
   GEORGE HERBERT,
      "The Flower"
      ll. 8–14 (1633)

One great gift of the holidays is the chance to connect with friends, who are some of the most enduring blooms in the gardens of our lives. This week I got a Christmas card from my friend Chris that reminded me of a happy afternoon last spring when we had tea in his garden. He introduced me to a new plant, yesterday, today, and tomorrow (*Brunfelsia pauciflora Floribunda*), whose blossoms are white the first day, lilac the next, and a deeper violet the third. Time brings its inevitable changes, but the joy of friendship extends through many yesterdays, todays, and tomorrows.

Chris was born in upstate New York, near Hyde Park, and has loved gardening all his life. He told me that when he was young and had no money to buy plants, he made a "friendship garden" with cuttings and plants his friends had given him from their gardens. He even made a chart showing which friend each plant represented. Chris's resourcefulness turned deficiency into opportunity. His garden became a living symbol, a tribute to his circle of friends, more beautiful than any plants from a garden supply store.

Working with the soil awakens our ingenuity. When I was young and without much money, I also made a garden of resourcefulness. In my first year of college, I lived with my parents in Colton, California, where one of my chores was pulling weeds and wild grasses on the hill behind our house. Although I enjoy working with plants, that weeding chore seemed pointless. It wasn't like weeding a garden, for nothing of any consequence was growing there, and every week or so the weeds and grass would reappear, requiring another weekend of work.

That year I was reading Albert Camus's *Myth of Sisyphus*, about a mythological character condemned to push a rock uphill for eternity. But unlike Sisyphus, who had to repeat the same task forever, I got an idea one day on my parents' hill. I needed to plant something that would take root, compete with the weeds, and end my repetitive labors. Since geraniums grow like weeds in southern California, they seemed a likely choice. Geraniums, or *Pelargonium domesticum*, *hortorum*, and *peltatum*, come in so many colors and varieties the hillside could be a beautiful sight and no longer a chore. I collected geranium cuttings on my daily drives to school, wrapped them in damp paper towels, and planted them on the hill when I came home. Whenever I drove past a house and spotted a new variety, I'd stop, knock at the door, and say, "Your geraniums are so beautiful. I wonder if I could have a cutting to take home to my mother." No one ever said no.

Before long I had a hillside planted with scarlet, orange, white, salmon, and pink geraniums, multicolored geraniums, and ivy geraniums in white, pink, and pale lilac. The blossoms brightened our hillside, and my mother actually did enjoy them, while I enjoyed being liberated from the repetitive weeding. Now on sunny weekends I could sit out on the deck with a glass of iced tea and whatever Shakespeare play I was reading, glancing up with a smile at the flowers on the hillside.

## Questions to Consider

THESE TWO STORIES and your own gardening experience attest to the power of resourcefulness. Applying this lesson to your inner garden, you can cultivate greater resourcefulness in your life this month by asking yourself these questions:

- What message does the friendship garden bring to me? How can I use my own resourcefulness to create new possibilities this holiday season?
- What is one holiday ritual I enjoy? How can I focus my attention on this and create a more personal celebration?
- What seasonal activities are stressors for me? How can I weed some of them out of my life to create more room for what really matters?

## Discipline and Dreams

THE DARKER DAYS of winter urge us to return to our roots like the perennials in our gardens, to find time amid the holiday commotion for our creative practice and the quiet pleasures of reading and contemplation. I've been enjoying my needlepoint and reading Thomas Jefferson's *Garden Book*, a book about Galileo Galilei, and the lives of some medieval saints. Manifesting their talents in different ways, all of these people led exemplary lives.

How did Jefferson keep his ideals alive in the face of heartbreak and disappointment? What gave Galileo the strength to carry on his work despite persecution by the Roman Catholic Inquisition? These people prevailed, overcoming adversity because they cultivated their inner lives. Saints Francis and Clare of Assisi turned from medieval Italy's warfare and materialism to live their spiritual ideals. Within the discipline of his holy rule, St. Francis found inspiration in nature. So deeply at peace was he that, according to legend, even the wild birds and beasts approached him as a friend. As their worlds shifted into revolutionary politics and cosmology, Jefferson and Galileo found comfort working in their gardens, cultivating and planting, feeling the earth in their hands, witnessing the restorative power of nature.

When the winds of change threaten to blow us off course or we're distracted by the noisy outside world, it takes discipline to keep our dreams

*How like a winter hath
my absence been
From thee, the pleasure of
the fleeting year!
What freezings have I
felt, what dark days
seen,
What old December's
bareness everywhere!*
WILLIAM
SHAKESPEARE,
sonnet 97,
ll. 1–4 (1609)

alive. The discipline of daily tasks is essential to cultivating the gardens within and around us. It is a spiritual exercise for which there is no immediate reward. We get little encouragement and often active interference from the people around us. Yet what creates an exemplary life is the courage and perseverance to keep cultivating our dreams, even when we seem to be making little progress.

A dream without discipline is a seed carried off by the wind. Without cultivation, the seed never takes root. Discipline without a dream makes of us mere plodders, working the soil without the courage to plant new seeds. Nothing grows. Discipline and dreams, soil and seeds. Both are essential: the daily work of discipline and the inner work, the quiet hidden work of seeds.

> *He shall be as a tree*
> *which planted grows*
> *By wat'ry streams, and in*
> *his season knows*
> *To yield his fruit, and his*
> *leaf shall not fall,*
> *And what he takes in*
> *hand shall prosper all.*
> JOHN MILTON,
> "Psalm I Done Into
> Verse,"
> ll. 7–10 (1653)

### Personal Exercise: Cultivating Discipline

TAKE SOME TIME by yourself this month to contemplate this lesson, asking yourself these questions.

- *Look around you:* Find someone you know who combines discipline and dreams. How does this person avoid distractions and keep cultivating the garden of his or her dreams? Is there something you can learn from this?
- Do you know someone with plenty of talent and dreams but no discipline, who plants seeds of new endeavors everywhere but lacks the patience and perseverance to cultivate them? Do you see this tendency in yourself? If so, what can you do about it? How can you add discipline to your dreams?
- Do you know someone else who never strays from the safety of a predictable routine to plant new seeds of possibilities? If you find yourself in this situation, resolve now to begin planting seeds for your dreams.
- *Look within you:* This holiday season, give yourself the gift of combining

discipline and dreams. Take an index card and write on it the dream you wish to realize, and one small but ongoing discipline you will undertake to cultivate it.

🐏 Put the card in a place where you will see it every morning and evening. For the next thirty days, beginning today, look at the card and affirm to yourself: "I have the discipline to cultivate my dreams."

## Garden Tasks

### Tending Cool-Weather Crops and Cultivating the Soil

*Fat crumbling Earth is fitter for the Plough, Putrid and loose above, and black below: . . . No Land for Seed like this, no Fields afford So large an Income for the Village Lord.*

John Dryden, *Virgil's Georgics*, II, ll. 280–81, 284–85 (1697)

If you live in a mild region, you may still be growing cool-weather crops that will need fertilizing, tending, and watering (if it doesn't rain much). Since we've had an unusually dry winter, I've been watering the plants still growing in my yard. In colder regions a deep watering before the soil freezes will help your plants make it through the winter.

This month I dug in more leaves along with the compost and manure in the north garden beds, where this spring I intend to plant Thomas Jefferson's favorite vegetable: peas. For now the plan is to enrich the soil. Earlier this month I bought some sulfate of ammonia at the hardware store and carefully added a small amount—one ounce per ten square feet—then spaded it in. The ammonium should break down into usable nitrogen by this spring. For now the garden beds are resting and building energy, which is what a lot of people and many animals do in the winter, springing back with new vitality when the days grow warmer.

### Endings and Beginnings

As more plants die back in our gardens, there are spent perennials and annuals to add to the compost pile. Endings and beginnings. The brilliant chrysanthemums of last month are now only a memory. I cut them back to six inches from the ground to let them rest and renew their energy for another season.

Before the first hard frost, it's time to dig the dahlias and store their tubers in a cool, dry place for the winter. Last year I did not do this, and only a few of my dahlias bloomed. Lifting the plants from the ground I could see why: their tubers had become overcrowded. I'll divide them this spring to give them lots of room to grow. For now I've been gradually lifting them out of the ground and tucking them in to rest in a quiet corner of my garage. If you live in a colder climate, you've probably stored away all your dahlias and other tender bulbs by now. You'll need to check them once a month to make sure the tubers are not rotting or drying out.

> . . . *let winter have his*
> *fee;*
> *Let a bleak paleness*
> *chalk the doore,*
> *So all within be livelier*
> *then before.*
>        GEORGE HERBERT,
>        "The Forerunners,"
>        ll. 34–36 (1633)

## Lifting and Storing Dahlias

WHEN THE FIRST frost withers their foliage and the plants go dormant, it's time to dig up your dahlia tubers and store them for the winter. Keep them in a dark, cool place and replant them in April or May, when the danger of frost has past.

- *Cut off the stems six inches from the ground.* Compost the spent foliage.
- *Using a spade or shovel, dig a circle* about two feet in diameter around the plant.
- *Carefully lift the tubers with your garden fork.* As you remove the tubers from the fork, shake off any loose soil.
- *Label the tubers by color and type and hang them upside down* for a few days in a cool, dry place using string or mesh bags.
- *When the tubers have dried, dust off any extra soil and examine them.* You can divide crowded tubers now or wait until spring. Dividing them in spring reduces the risk of rot or shriveling in storage.

* *Store the tubers stem side up* in a box or paper bag filled with sand, sawdust, dry peat moss, or crumpled newspaper. Place the container in a cool (40 to 45 degrees) dry, frost-free place. (I keep mine in my garage.)
* *Inspect the tubers once a month.* Check for shriveling. If the tubers are drying out, mist them with water or dampen their storage material. Check for rot, discarding any infected tubers. Dust the others nearby with fungicide to prevent further contamination.
* *In the spring, when the danger of frost has passed,* divide overcrowded clumps with a sharp knife, leaving a portion of stalk attached to each tuber. Dust the cuts with fungicide. Place the tubers in moist sand for two to three weeks to encourage sprouting. Then plant them in your garden in well-cultivated soil rich in organic matter, in holes one foot deep and at least one foot across, providing plenty of space (four to five feet apart for large dahlias, one to two feet apart for smaller ones).

## Late Harvests and Trimming Evergreens

IF YOUR AREA has mild winters, you may have some late vegetables to harvest. The main vegetable harvest I have now is tomatoes—from my windowsill. The green tomatoes I brought inside last month have been providing us with homegrown tomatoes for our salads for weeks. The sweet basil has died back, but the hardier rosemary, sage, marjoram, and thyme are still doing reasonably well outside, seasoning our winter soups and roasted dishes. A few green onions are still growing in a planter near our front door.

My other harvest this month has been leaves, which I now see as a valuable resource, not something to be cleaned up and swept away. Raking up the fallen leaves in my yard and bagging them to make leaf mold is a wonderful way to recycle, returning nature's nutrients to the soil.

December is a good time to trim evergreens, to shape them if they've gotten unruly and to collect fragrant boughs for holiday decorations. I once made a small Christmas tree from evergreen boughs trimmed from a giant pine beside my apartment building. I bound together the ends of the boughs with wire to form a trunk, then set the evergreens in a pot of water and decorated my "tree" with miniature ornaments. Its fragrance filled my tiny apartment and brought back memories of many a Christmas past. If you don't want to spend hours making wreaths or more elaborate decorations, you can still place evergreen boughs on your mantle or on top of a china cabinet, adding a red bow, candy canes, or some ornaments if you like.

### Garden Gifts

SOME OF MY favorite holiday gifts come from the garden. Each year my friend Tracey bakes loaves of spicy persimmon bread to give to her friends. I've given people bottles of dried homegrown herbs or lavender sachets. I also like to give my friends small pots of herbs I've grown from cuttings or picked up at a nursery. Rosemary, thyme, bay, and other culinary herbs add welcome greenery to kitchen windowsills and fresh flavor to winter meals.

> *But peaceful was the night*
> *Wherein the Prince of light*
> *His reign of peace upon the earth began.*
> JOHN MILTON,
> "On the Morning of Christ's Nativity,"
> ll. 61–63
> (1629)

Another favorite gift plant is *Aloe vera,* a member of the lily family legendary for its healing properties. Native to Africa and the Mediterranean, healing aloes are mentioned in the Bible. The Romans and Egyptians treated wounds with aloe, and Cleopatra reportedly used it in her cosmetics. In Western Europe this remarkable plant has been used since the Renaissance to heal wounds and sores, soothe itching and skin irritations, and treat all kinds of internal ailments.[6]

Today aloe vera gel is sold in pharmacies to treat sunburn, chapped or irritated skin, minor cuts, and burns. It's also added to many cosmetics. I keep an aloe vera plant in my kitchen window so I can clip one of its leaves and apply the fresh gel to minor cuts and burns. I've also given many aloe veras to family and friends. They are attractive, low-maintenance houseplants that require little water. Tied up with red ribbons, they make simple, thoughtful holiday gifts.

## Winter Care for Tender Plants, Livestock, and Tools

THOMAS TUSSER'S ADVICE for Renaissance gardeners this month was to sharpen their tools, put their cattle inside for the winter, take care of their horses, and cover their strawberries, rosemary, and other tender plants to protect them from frost.[7] If you live in zones 1 through 7, by now you've already taken steps to protect your plants from freezing weather. In zones 8 and 9 you'll need to listen for frost warnings and cover tender plants at night. Last December I spent many nights putting blankets on our lemon trees, but the weather this year has been milder. Rosemary survives outdoors in our winter weather, and my strawberries are protected in the inner courtyard. Few of us need to heed Tusser's advice about cows and horses, but we all need to take care of our tools.

*There is a balsome, or
  indeed a bloud,
Dropping from heav'n,
  which doth both cleanse
  and close
All sorts of wounds; of
  such strange force it is.
Seek out this All-heal,
  and seek no repose,
Untill thou finde and use
  it to thy good:
Then bring thy gift, and
  let thy hymne be this.*
  GEORGE HERBERT,
  "An Offering,"
  ll. 19–24 (1633)

### Cleaning and Storing Tools

WHEN YOU'VE FINISHED most of your active work in the garden this season, remember to clean and store your tools so they'll serve you well for many garden years.

- *Clean off any caked-on soil* from shovels, spades, trowels, and garden forks.
- *Sharpen any tools that need it.* Use a file or whetstone to sharpen dull edges. Take your pruning shears and lawn mower to the hardware store to have them sharpened professionally.
- *Clean off any signs of rust, and wipe tools with an oily rag.* I like to clean my tools with fine steel wool, then spray them with olive oil cooking spray and wipe away the excess. Don't forget to clean and oil wooden handles, too.

- *Store tools in a shed or in a corner of your garage.* Take time to organize your tools, hanging them neatly in rows or storing long ones in a barrel. Note any that need repair or replacement and take action now to be ready for winter pruning and spring planting.

### Dreaming of Tomorrow's Gardens

RECENTLY THE THOMPSON & Morgan seed catalog came in the mail—two hundred luscious pages of flowers, herbs, and vegetables. In December, when sunset comes early and time in the garden is scarce, we can still enjoy gardening in our imaginations, looking at catalogs and dreaming of another garden year.

I spent yesterday evening with a cup of spice tea and the seed catalog, considering all the colorful possibilities for next year's garden. I chose garden poppies (*Eschscholzia californica*), in Thai silk colors of scarlet, bronze, and rose, to join the golden California poppies that reseed themselves annually in my front gardens, and a packet of *viola* yesterday, today, and tomorrow, which reminded me of the larger plant I'd admired in Chris's garden. I also selected an exotic mixture of nasturtiums (*Tropaeolum majus*) in burgundy, tangerine, and cream, and some pansies (*Viola wittrockiania*) in deep velvet red, purple, and gold. Then there were the vegetables: two varieties of Thomas Jefferson's favorite early peas (*Pisum sativum*), plus some sugar snap peas, some Kentucky blue climbing beans (*Phaseolus vulgaris*), and a variety of miniature sweet bell peppers (*Capsicum annum*). Along with my usual tomatoes and greens, these should create a delightful kitchen garden next year.

**Seed Catalogs**
Looking through seed catalogs is an enjoyable task in December. Here are some good ones to try:

- **Burpee Seed Co.** *300 Park Avenue, Warminster PA 18991 (800-888-1447) www.burpee.com*
- **Bountiful Gardens** *(for organic heirloom seeds), 5798 Ridgewood Road, Willits CA 95490 (707-459-6410) www. bountifulgardens.org*
- **Nichols Garden Nursery,** *1190 Pacific Highway, Albany OR 97321 (541-928-9280) www.gardennursery.com*

Take some time this month to look through your garden catalogs, recall the gardens of your past, and let your imagination dream of gardens yet to come.

## GARDEN REFLECTION

### *Nature's Garden, Nature's Cycles*

NEAR THE END of the month, Bob and I spent a day hiking in Big Basin Redwoods State Park. We walked through the quiet forest, surrounded by the fragrance of evergreens, dwarfed by the redwood trees. *Sequoia sempervirens*, giant sequoias, their massive trunks reaching for the sky, are the oldest living things on the planet, some of them two to three thousand years old. Some of these trees were growing here long before Jefferson cultivated his garden and wrote the Declaration of Independence; before the first Europeans set foot on this continent; before a decree went out from Caesar Augustus and Joseph of Nazareth journeyed to Bethlehem with his wife, Mary, who bore her firstborn son and laid him in a manger.

Reaching back through the mists of time, spanning epochs, these epic trees have endured. Their trunks struck by lightning, battered and charred, many of them still survive, spreading their graceful branches to the sky. When a giant redwood does fall, its mighty trunk sprawls like a leviathan across the land. Young green shoots often emerge from the roots, and the cycle of life continues. In other, quieter cycles the needles and cones fall to the ground, the needles slowly turning to compost, enriching the soil to nurture more growth. The cones provide the seeds of future giants that will stand here a thousand years from now.

- **Park Seed Co.,**
  1 Parkton Avenue
  Greenwood, SC 29647
  (800-845-3369)
  www.parkseed.com
- **Shepherds Seeds,**
  30 Irene Street,
  Torrington CT 06790
  (860-482-3638)
  www.shepherdseeds.com
- **Thompson & Morgan,**
  **Inc.,** P.O. Box 1308,
  Jackson NJ 08527-0308
  (800-274-7333) www.
  thompson-morgan.com
- **Wayside Gardens,**
  1 Garden Lane,
  Hodges SC 29695-0001
  (800-845-1124) www.
  waysidegardens.com

Whether written high above our heads in the lives of redwood trees or in the smaller fonts of our own gardens, the lessons of nature encircle and include us all. As this final month of the calendar year leads to a new annual cycle, the season of winter leads to another spring. In our lives in so many ways the end touches the beginning, as the darkness of winter affirms the renewal of life.

# THIRTEEN

## January:
## Winter Pruning and
## Garden Planning

For never-resting time leads summer on
To hideous winter, and confounds him there,
Sap checked with frost, and lusty leaves quite gone,
Beauty o'ersnowed, and bareness everywhere.

WILLIAM SHAKESPEARE,
sonnet 5
ll. 5–8 (1609)

JANUARY BRINGS US into deepest winter. Bare branches reach
up to a cold gray sky. Our days are stripped of holiday color as we
pack away our decorations, clearing the debris of wrapping
paper, cards, and boxes.

The garden brings us valuable lessons in this cold, dark season. Under
the winter sky and bare branches, animals hibernate and much of nature
sleeps. This month calls us to slow down, take stock, and center ourselves
by returning to our roots.

From the Middle Ages through the Renaissance, the first week in January marked a major transition. Holiday festivities continued from New
Year's to Epiphany on January 6. Marking the visit of the three kings who

followed the star to Bethlehem, this final feast of Christmas or Twelfth Night was celebrated with festive cakes and revels.

The next day the holidays were over. People put away their finery and went back to work. On January 7, St. Distaff's Day, women returned to their spinning and domestic routine. The following Monday, Plough Monday, or as soon as weather permitted, farmers went back to their fields, plowing the soil and pruning their dormant trees and vines.[1]

After all the holiday opulence, January returns us to life's routine, a change that some regret and others, depleted by too much celebration, welcome. As our gardens remind us, alternating patterns of darkness and light, action and contemplation are essential to nature. Their alternation restores our balance, for, as Shakespeare realized, too much of either extreme will exhaust us:

> If all the year were playing holidays,
> To sport would be as tedious as to work.[2]

In Shakespeare's time January was proverbially the coldest month and January 13, St. Hilary's Day, the coldest day. Because of inclement weather, people spent much of January indoors by their hearths, trying to keep warm.

On January 20, the Eve of St. Agnes, young maidens would fast all day, then eat a salted hard-boiled egg or herring before going to bed, hoping for dreams of their future love.[3] In our gardens and lives today, January is a time to begin turning our dreams into action, pruning away the past and planning for tomorrow, as we enjoy the quiet beauty of winter.

*Of Twelf-tide Cakes, of*
*   Pease, and Beanes*
*Wherewith ye make those*
*   merry Sceanes,*
*When as ye chuse your*
*   King and Queen,*
*And cry out, Hey, for our*
*   town green.*
*Of Ash-heapes, in the*
*   which ye use*
*Husbands and Wives by*
*   streakes to chuse:*
*Of crackling Laurell,*
*   which fore-sounds,*
*A Plentious harvest to*
*   your grounds:*
*Of these, and such like*
*   things, for shift,*
*We send in stead of New-*
*   yeares gift.*
      ROBERT HERRICK,
   "A New-yeares gift sent
   to Sir Simeon Steward"
      ll. 17–26 (1648)

## GARDEN GROWTH

*Give St. Distaffe all the*
*right,*
*Then bid Christmas sport*
*good night.*
*And next morrow, every*
*one*
*To his own vocation.*
ROBERT HERRICK,
"St. Distaffs Day,"
ll. 11–14 (1648)

*I freeze, I freeze, and*
*nothing dwels*
*In me but Snow and*
*ysicles.*
*For pitties sake give your*
*advice*
*To melt this snow, and*
*thaw this ice.*
ROBERT HERRICK,
"The Frozen Heart,"
ll. 1–4 (1648)

IN A GARDEN every season brings its own kind of growth. We can benefit from this month's spare simplicity by becoming more mindful of the subtle signs of life in our midst. January's bleak canvas challenges us to notice details. In my garden today I saw two brown squirrels scampering across the branch of an oak tree high overhead. A flock of Canada geese swept across the silver-gray sky. Looking down at the sodden cobblestone path, I noticed one small sprig of white alyssum amid the fallen leaves.

The bare branches in my backyard are a testament to winter, shadows of stark emptiness where fragrant wisteria once festooned the redwood arbor. All around me are emblems of beauty, loss, and renewal. In the front yard the trees are bare and most of the annuals are gone. The summer dahlias, blasted by the frost, have turned to clumps of brown tubers stored in my garage. The roses of summer have shed their petals and leaves, dying back to winter dormancy, while at their feet early bulbs are raising hopeful green shoots from deep within the earth. In our gardens and our lives, cycles of endings and beginnings repeat nature's essential rhythms within and around us.

This month, when weather permits, we can prune our dormant trees and shrubs. Like men and women in bygone days, we can stay by our hearths on cold winter nights, putting last month's garden dreams into action as we lay out our plans for the year ahead. We can also create new traditions of planning and pruning in the gardens of our lives.

## GARDENING AS SPIRITUAL PRACTICE

### *Moving from Dreams to Action Plans*

LAST MONTH I dreamed of future gardens, ordering seeds for new flowers and vegetables. This month I'm taking the next step: deciding where to sow them, determining how these new plants will fit into my garden design.

Last month you chose a dream to work on, writing it on an index card along with one discipline to begin moving toward it. This month you can take the next step: turning your dream into an action plan.

Personal growth, like growth in a garden, happens in stages. Years ago I watched Bob begin his campaign to stop smoking, affirming his dream for a healthier life. As a psychologist he knew that the first step in changing any behavior is awareness. So for a while he simply recorded how many cigarettes he smoked, keeping count on an index card he carried in his pocket. This was his disciplined action to begin moving toward his dream.

The discipline itself motivated him to cut down. He gradually went from two packs of cigarettes a day to one, then a half pack, then a handful. Around the same time he began running as his exercise practice. As his smoking decreased, his running improved. He made himself a plan of action, and his forward momentum increased as one step led to another. He stopped smoking, began eating more healthful foods, and ran his first marathon in 1987, followed by six more.

Bob's ability to take action and accomplish his dreams impressed me then as it does now. I have watched him use the same momentum at the university when he chaired his department and now as he does his research on the brain. Dreams, discipline, and deliberate action—it's a formula for success that reaches back through centuries and across continents. The conclusion of Milton's *Paradise Lost* affirms this vision of human possibility: "by small, Accomplishing great things." The ancient Chinese book of wisdom the *Tao Te Ching* tells us, "The journey of a thousand miles begins with a single step." Writers, scientists, statesmen, and leaders in all fields have known the power released when we take one small step, then another, moving forward with wisdom and confidence, as we link our dreams to deliberate action.[4]

## Personal Exercise: Discipline, Dreams, and Deliberate Action

THIS MONTH, YOU can build positive momentum by creating an action plan for yourself. Take out your card from last month with the dream and the discipline you chose. Give yourself a quiet hour and let the following questions guide you in creating your plan:

- Where has the discipline brought you? Have you made progress toward your dream?
- In what ways have you progressed? What do you need to do now?
- What new possibilities have arisen?
  - How can you follow up on them?
  - Write down these steps.
- What (if any) roadblocks have arisen?
  - What actions can you take to overcome them?
  - Write down these steps.
- Make yourself an action plan for the month ahead.
  - Write down specific actions with dates and times to accomplish them.
  - Put them on your calendar.
- Record the main points of your action plan on an index card with the affirmation "I have the discipline and deliberate action to achieve my dreams."
- Look at this card at least once a day.
  - Smile as you feel yourself filled with new, positive momentum.

## Personal Exercise: Inner Garden Design

JANUARY WAS NAMED for the Roman god Janus, guardian of new beginnings, doors, and gateways. He was portrayed with two faces, one looking forward, the other looking back. January has always been a time for long-range planning: learning from the past as we move into the future. Before you review your garden design this month, take some time to review the design of your life.

- Close your door, unplug the phone, and give yourself at least an hour and a half when you won't be disturbed.
- Take a sheet of paper and make headings for the five vital areas of your life: work, play, love, health, and spirit. These areas represent different things to different people.
  - Most of us associate work with our jobs, but some artists have "day jobs" to pay the rent that are quite different from their life's work.
  - All of us need play, which to some may be a favorite sport, to others a favorite hobby, games with our children, or time with our pets.
  - Love takes many forms: romance, family, friendship.
  - Health is our physical and emotional well-being. Maintaining it means being mindful of our needs.
  - Spirit involves a range of possibilities that nurture greater joy, depth, and meaning in our lives. For some this means orthodox religious practice, for others quiet walks in nature. For centuries gardening has nurtured many people's spirits.
- For each of the five areas, ask yourself: Did this area of my life flourish last year or disappoint me? Write down your answers and any strengths or weaknesses. Spend ten to fifteen minutes on each area.
- When you have finished, look at the five areas and ask yourself: Does this design meet my needs, or is something valuable missing?
- If something is missing, what is it? Write it down in the area where it belongs and circle it.
- For each circled need, think of *one simple thing* you can do to begin meeting it. If your health is weak because you haven't found time to exercise, think of one thing you can do—taking a walk at lunchtime, doing stretching exercises in the morning to strengthen your back, signing up for a fitness class, or something else that you feel motivated to do. If you need to nurture your spirit more, what can you do? Begin

each day with prayer or meditation? Attend church? Take solitary walks in your local park? Choose only *one* action per area. Write the action down and circle it.

✎ Now make an action plan for January. Take out your calendar and schedule *when* you'll do these things. Then set your design aside for later reference.

Designing your life this way is far more effective than writing a list of New Year's resolutions. By scheduling your choices, you'll be adding action to your plan, ensuring greater success by putting your needs on the calendar first.

Personal planning is something many of us take for granted, but the individualism and agency we enjoy today came with the improved technology of the Renaissance. In the Middle Ages most people rose with the sun, worked all day, and went to bed early.

Their days were marked by the sun's rounds; their lives by agrarian rhythms. More precise measurement of time began in medieval monasteries, where the Rule of St. Benedict divided the day into set times for work, study, and prayer. Living and working communally, the monks moved through days marked by bells announcing the canonical hours of vigils, lauds, prime, terce, sext, none, vespers, and compline. Each monastery had its rudimentary clock with a sacristan assigned to toll the hours. By the twelfth and thirteenth centuries, medieval towns kept time together as town clocks tolled the hours and town criers rang out the curfew each night. But by the late fifteenth century watches were invented and timekeeping was individualized. Renaissance men and women could consult their watches and plan ahead, keeping time for themselves, designing their lives with greater agency.[5]

*In winter at nine, and in sommer at ten*
*To bed after supper both maidens and men.*
*In winter at five a clocke, servant arise,*
*In sommer at foure is verie good guise.*
THOMAS TUSSER,
*Five Hundred Points of Good Husbandry*
(1580)

## *Personal Pruning*

JANUARY IS TRADITIONALLY a time for pruning, cutting back, eliminating those things that no longer serve us. Last Friday as I left for work, my neighbor Rhonda and a friend were cutting down the invasive wisteria vine growing between our yards. The vine had been there since I moved in, but it had never blossomed. Instead, it twined through the fence like a python, crushing the latticework, snaking over the gate and onto the roof. Last year we cut it back repeatedly, but it came back more determined than ever. Later that day, when I returned from campus, the massive vine was gone, the redwood fence and gates beautifully clear and serene.

In our lives as in our gardens, too much of something can become invasive, upsetting our balance and obscuring the beauty of our lives. Since the Middle Ages the increasing precision of clocks has been a mixed blessing, moving us by mechanical rhythms instead of natural cycles. By the late Middle Ages, not only monks but people in urban areas began measuring their days by clocks and bells. Minute hands were added to clocks in the fifteenth century, half hour and quarter hour markings in the seventeenth century, and second hands in the 1690s.[6] As cities became busy centers with people rushing to appointments, their lives controlled by the instruments they had made, people in the Renaissance developed a passion for gardens. While tending their herbs and flowers, they enjoyed quiet contemplation. Returning to nature's cycles, they regained their sense of balance.

Gardens also taught our Renaissance ancestors valuable lessons. Pruning, an important winter task, affirmed the importance of order and discipline. In Shakespeare's *Richard II* three palace gardeners discuss King Richard's inability to maintain order in his country. The head gardener tells his assistants to

> cut off the heads of too-fast growing sprays
> That look too lofty in our commonwealth.
> All must be even in our government.

> You thus employed, I will go root away
> The noisome weeds which without profit suck
> The soil's fertility from wholesome flowers.

"What a pity," laments the head gardener, that King Richard "had not so trimmed and dressed his land/ As we this garden."[7]

Seventeenth-century devotional writers related pruning to spiritual growth. As regular pruning strengthens our fruit trees and roses, pruning was also the means by which God, the Divine Gardener, strengthened people's souls. George Herbert's poem, "Paradise" cleverly prunes the final word of each line:

> I blesse thee, Lord, because I GROW
> Among thy trees, which in a    ROW
> To thee both fruit and order      OW.
>
> What open force, or hidden CHARM
> Can blast my fruit or bring me HARM,
> While the inclosure is thine      ARM?
>
> Inclose me still for fear I    START
> Be to me rather sharp and     TART
> Then let me want thy hand and ART.
>
> When thou dost greater judgements SPARE,
> And with thy knife but prune and     PARE,
> Ev'n fruitfull trees more fruitfull      ARE.
>
> Such sharpnes shows the sweetest FREND:
> Such cuttings rather heal then     REND:
> And such beginnings touch their     END.[8]

## *Personal Exercise: Pruning in Your Life's Garden*

LIKE OUR RENAISSANCE and seventeenth-century predecessors, we can apply the lesson of pruning to our lives. You can begin this process by taking out your life's design sheet, reviewing the five areas of work, play, love, health, and spirit, and asking yourself these questions.

- Are any areas so overgrown that they're crowding out the others?
  - If so, how can you cut back in these areas?
  - If you've been saying yes to too many projects at work, which ones can you cut back and delegate to allow more time for play, love, health, and spirit? Sometimes we don't delegate routine work, but run ourselves down doing it all ourselves. This unhealthy habit needs pruning!
- Are some areas of your life so tangled with old growth that they lack vitality? Like lavender and roses in the gardens, these areas will spring forth with new vigor if you prune away old routines to make room for something new.
  - If a project or relationship has become enmeshed in dull routine, can you and the other person break out of the pattern and try something new? I'd been coauthoring a book with my friend Tina and making little progress. Last winter we discussed our frustration and decided to write a shorter book. We both felt a new burst of enthusiasm.

### *More Personal Pruning*

ONCE YOU'VE PRUNED your life's design, keep pruning in your daily activities. Challenge yourself this month with the question, "What can I prune in my life today?" This week I've found myself pruning:

- energy-draining activities from my schedule by asking, "Do I really need to do this?" Often the answer is no. Once I started pruning, I've found ways to delegate and consolidate many tasks.

- requests at work to add to my packed schedule. It's easier to say, "No, I cannot take on any new projects right now," when I see this as pruning.
- unnecessary objects from my briefcase and handbag. By taking out the junk, I've lightened my load and made things easier to find.

## Economic Pruning

PRUNING OFTEN MEANS cutting away one thing to make room for something else. Years ago, while saving to buy a new car, I realized I'd been spending almost eight dollars a day for lunch at the faculty club. By packing my own lunches—and bringing healthful foods I enjoyed—I easily saved over one hundred dollars a month.

This economic pruning brought me added dividends. During lunch I saved time, got more exercise, and expanded my day. On warm days I'd eat outside, then walk around campus. Or I'd eat my lunch at my desk, then walk over to the library, do an errand, write a letter, or spend some quiet time with a good book. I've had my new car for years now, but I still bring my lunch to work most of the time because I enjoy the freedom and flexibility.

## Attitudinal Pruning

A MORE SUBTLE pruning involves our thoughts and attitudes, clearing away excessive worry and complaint. Continuing to worry about a problem or ruminating leads to a negative, defeatist attitude known as "learned helplessness." The more we worry about a problem without taking action, the more helpless and hopeless we feel. Ruminating is actually hazardous to our health. Psychologists have found a strong correlation between rumination, and clinical depression.[9]

Another habit to prune away is chronic complaining. Look around at the people you know. Chances are the constant complainers spend most of their time and energy doing just that, rarely getting around to solving the problem. The most productive people I know are upbeat, centered, focused on their own priorities. When they see a problem they take positive action: either solving it themselves or referring it to someone who can.

## Personal Exercise: Taking Positive Action

THE NEXT TIME you find yourself facing a problem,

- Ask yourself, "What can I do about it?"
- If you can do something, do it. Take positive action.
- If you can't solve it, find someone who can. Delegate or ask for help.
- Don't ruminate or get caught in a negative spiral of worry or complaint. Remember where you're headed and keep moving forward.

## GARDEN TASKS

### Pruning Trees, Shrubs, and Roses

WITH MANY PLANTS dormant this month, our garden tasks diminish. Pruning is the main task, along with a little garden maintenance.

You'll want to prune your deciduous trees and shrubs while they're still dormant. In my climatic zone spring can come early, so we do lots of pruning in January. In the middle of the month, Bob and I pruned our hybrid tea roses.

Pruning your roses for the first time can be a shock, for the canes are cut back to about half, leaving the bushes looking quite decimated. If you're reluctant to prune your roses because you don't know how, check the garden section of your newspaper for local classes and demonstrations. When Bob moved into his house in San Jose, he went to a pruning workshop at the municipal rose garden, then came home and pruned his roses with confidence.

*Much labour is requir'd in Trees, to tame Their wild disorder, and in ranks reclaim.*
JOHN DRYDEN,
*Virgil's Georgics*
II, ll. 85–86 (1697)

Pruning imposes discipline in your garden, removing dead and overgrown branches, cutting back old growth to make way for new. After Bob and I pruned our rosebushes, they looked scaled down, simplified, ready for a new season of growth. Then we walked around the yard with our loppers and

*But let thy Hand supply
  the Pruning-knife;
And crop luxuriant
  Straglers, nor be loath
To strip the Branches of
  their leafy Growth:
But when the rooted
  Vines, with steady Hold,
Can clasp their Elms,
  their Husbandmen be
  bold
To lop the disobedient
  Boughs, that stray'd
Beyond their Ranks: let
  crooked Steel invade
The lawless Troops, which
  Discipline disclaim,
And their superfluous
  Growth with Rigour
  tame.*

JOHN DRYDEN,
  *Virgil's Georgics*
II, ll. 502–10 (1697)

garden shears, pruning the withered hydrangea bushes and cutting back branches from the avocado tree that were growing across our north garden path. There is still more to do, but I'm enjoying all the pruning tasks this month. They give me a feeling of agency, as I work with nature to create greater order in this small part of the world.

## Pruning Tips

- Prune deciduous plants when they're dormant—in late winter or early spring.
- Prune spring-blooming shrubs after they've bloomed; if you prune them now, you'll cut off their buds and rob yourself of spring blossoms.
- To encourage denser growth, prune shrubs to buds facing inward.
- To open up a plant, prune to outward-facing buds. (Roses are best pruned this way.)

## Pruning Your Roses

SOME PEOPLE PRUNE their rosebushes in winter. Gardeners in colder regions wait until early spring, when they remove the bushes' winter protection. But be sure to prune your roses while they're still dormant.

- Cut back the canes of hybrid tea roses to about half.
- Remove all dead or diseased wood.
- Cut out any thin spindly canes or canes that cross and crowd each other.
- Think of each bush as a basket shape. Cut out any canes growing inside to leave the center open and give your roses breathing room.
- Cut the canes above an outward-facing bud or above a bud in the direction you want your rosebush to grow. If you cannot find a bud, cut the cane above an outward-facing leaf with five leaflets.

All plants need pruning periodically. Pruning gives plants better shape, gets rid of suckers coming up from the roots, eliminates deadwood, overgrown branches, and faded flowers. Deadheading roses and other flowering plants during the growing season is another form of pruning that encourages further blooms.

### Garden Maintenance

WHEN IT'S NOT raining I've been pulling weeds and clearing away fallen leaves, spent annuals, and other garden debris. I found some emerging bulb foliage damaged by snails or slugs, so I scattered snail bait around the garden. I also dug in some bonemeal to feed the spring bulbs.

Depending on the weather in your region, you may need to knock snow off fragile shrubs, check your winter mulch, or cover your tender plants to protect them from frost. I've been listening for frost warnings and watching the sky—clear, cloudless nights are the coldest—so I'll know when to cover my citrus trees.

Indoors, I've been checking the dahlia tubers at least once a month. If they start to shrivel and dry out, I'll mist them with water. If they show signs of rot, I'll cut off the affected parts and dust the tubers with sulfur. Another indoor task is caring for tools. I just cleaned our pruning tools before putting them away.

### Garden Exercise: Planning Ahead by Looking Back

THIS MONTH WE can plan our gardens of tomorrow by first reflecting on last year's gardens and making some notes. If the weather is not too stormy, I like to walk through my garden and look at each of the beds. In your garden or your imagination, revisit last year's garden, going from one bed to another, and ask yourself:

- Which plants grew well here last year: in the early spring, summer, and fall?
- What did I find disappointing?

- If something didn't grow well, can you think of some reasons why:
  - Was the sun too hot or was it too shady for this kind of plant?
  - Was the soil too acid? Too alkaline? Lacking in nutrients?
  - Did the plant get enough water? Too much?
  - What can you do this year to help your garden beds grow better?

Looking back on my garden, I remembered the roses and spring bulbs—all star performers that had the right conditions of soil, sunshine, and water to excel. The tomatoes, too, were impressive, especially after I began feeding them regularly. But the dahlias and green beans were disappointing. Only a few of the dahlias bloomed, and the beans were a continual challenge.

## Making New Garden Plans

I'VE INCORPORATED LAST year's lessons into this year's garden plan. This summer I will feed and water all the dahlias more. Before I replant them this spring, I'll divide them, for when I dug them up I found their tubers grown together in massive clumps. The green beans of last spring had many setbacks: snails destroyed the French beans, sown in early spring. Later sowings lacked sufficient moisture to germinate. A final crop was started late, and their soil lacked nutrients.

This year I've been preparing the soil, adding lots of organic matter: compost and horse manure, as well as sulfate of ammonia for nitrogen. Later this month I'll dig the beds again, then give them a final digging before sowing an early crop of garden peas.

Gardening is part art, part science. Germination and soil conditions follow scientific rules. The more we know them, the better our gardens will grow.

## Reviewing Your Garden Design

ANOTHER ASPECT OF garden planning is design: combining artistic principles of composition and color with your intuition and personal taste. In one sense our gardens are always growing. Even in winter dormant plants and bulbs are preparing to spring to life. And in the garden of your imagination, new insights, new possibilities will emerge.

January, when our gardens are stripped down to the bones, is a good time to consider garden design. I like to take slow walks around my garden looking for form—horizontal and vertical shapes, contrasts between the garden beds and the brick or cobblestone paths, the vertical lines of the garden fence, tree trunks, and arbors, and the horizontal spaces in between.

In designing our gardens we cultivate the art of seeing, looking for shape and contrast. Winter helps with this process by eliminating distracting foliage and color. Design is more apparent in a winter garden, as it is in black-and-white photography. We can train ourselves to become more aware of form and composition. Some people draw their gardens on graph paper, a good idea if you are laying out a new garden or wondering how new plantings will harmonize. There are even computer programs for garden design. But since my garden was beautifully designed by the artist who planted it, all I do is walk around, making brief sketches and notes.

### Garden Exercise: Exploring Your Garden Design

TAKE A SLOW walk around your garden one day this month when weather permits. Bring along a pencil and your garden notebook or some index cards.

* *Look at your boundaries*—your garden fences and walls. How well do they harmonize with the plantings within them?
* *Consider the form of your garden.* Straight lines and geometric shapes are more formal; curving designs more casual. What do you have? Is this what you prefer?
* *Look at each of your garden beds and borders.* How well did their plantings harmonize last year? Would you like more height? More annual color when spring bulbs die down?

- *Look at the shapes of trees and shrubs.* How well do they harmonize with the rest of your garden? Do they need pruning?
- *Consider height.* When planting a new border or adding new plants, put the tallest plants in back, shorter ones in the middle, and the shortest ones in front so that all are visible. Could any of your garden use more height?
- *Look at your garden paths.* How well do they harmonize—in shape, color, texture, and function? Is the brick path too narrow? Are some of the cobblestones loose?
- *Look at your garden seating.* Are the decks functional as well as attractive? Could you add a garden bench? Do you use the chairs and table on your sunny patio, or would you be more comfortable in a shady arbor?
- Make notes to yourself about all these things and keep your eyes open for new ideas. Look at pictures in garden books and magazines. Sketch out your new design, make notes in your garden journal, or chart your new garden on graph paper.

Our gardens grow through time as well as space. Last spring the iris garden in my front yard was resplendent with color. For two months there was a changing display from the early crocuses, daffodils, and grape hyacinths to tulips, anemones, freesia, and irises of all imaginable colors. But then the show stopped, leaving only memories and rows of fading foliage. This fall I planted liatris and allium bulbs to give the garden bed more height and color from late spring through summer.

A garden is always a work in progress. As the artist you expand your garden design over time: planting bulbs and annuals to provide continual interest from spring to early autumn, adding new fruits, flowers, and vegetables to nature's evolving design.

## Ordering New Seeds and Plants

THE SEEDS I ordered last month just arrived: packets of pansies, nasturtiums, poppies, and violas; garden peas, green beans, and red bell peppers. After studying your garden design, you may want to order new seeds and plants to add to next year's garden. This month I got out my catalogs and ordered a bay laurel tree (*Laurus nobilis*) and two dwarf blueberry bushes to plant in pots in our front courtyard, some seed potatoes for our south kitchen garden, and some creeping thyme seed to sow between the cobblestones along our front garden walk. I wanted to add interest to our entryway, and I like the idea of fragrant thyme leading from the herb garden to our front gate.

When I saw them in the catalog, the small Epicure potatoes, a variety over a hundred years old, sounded too delicious to pass up. The laurel will give us bay leaves for cooking and make a beautiful topiary tree. Blueberries are my husband Bob's favorite fruit, a reminder of childhood summers in the Catskill Mountains, where blueberries grew wild and his aunt baked unforgettable blueberry pies. These blueberry bushes are *Vaccinium corymbosum* Sunshine Blue, ideal for growing in pots or small spaces and well suited to our climatic zone because they don't require much winter chill.

## Designing with Color and Light

AS GARDENERS WE know how each season brings different qualities of light. The gray skies of January give a light much different from the blazing sun of hot July. In my garden this month even the evergreens are muted, and many dark greens and browns recede under the steel gray sky.

Each day has its cycles, from the soft morning light through the glare of high noon to the late afternoon light that brings out the deepest greens and blues. You're familiar with your garden's microclimates: the sunny southern beds where tomatoes thrive; the checkered shade preferred by fuchsias, impatiens, and most salad greens. But now, while most of your garden is still asleep, you can cultivate a greater awareness of color.

Part of color awareness involves science; the rest is art. In 1666, working with a prism, Sir Isaac Newton broke white light into what we know as the color spectrum: red, orange, yellow, green, blue, indigo, and violet— seven colors he related to the seven notes of the musical scale. Mixed in different tones and intensities, these colors bring warmth and vitality to our gardens and our lives. Even in the wintertime the crystal paperweight on my windowsill catches the afternoon light, sending rainbow patterns dancing across the walls.

This winter I picked up a color wheel at an art supply store to expand my awareness of color. Colors next to each other on the wheel—green and blue, red and orange, orange and yellow—are known as harmonious. Stronger contrasts come with complementary colors, opposite each other on the wheel: orange and blue, yellow and violet, red and green. This summer I planted blue lobelia among the golden California poppies in my front garden because I liked the contrast.

You can create subtler effects in your garden by using pastel shades, like pale yellow and lilac, or by blending your colored blossoms with white. Dramatic color effects are also possible with monochromatic gardens—entire beds in one color or its gradations. White gardens can be quite magical, and I've seen beautiful beds in hues from rose to pink. But I love color too much to choose only one at a time.

Another way to create harmonious patterns is weaving similar colors throughout a garden bed. My review of garden design has inspired me to add more violets and lilac shades to the iris garden next year. This spring I'll plant society garlic (*Tulbaghia violacea*) like Rhonda's to add height and conceal the fading iris foliage. Its rose-violet blossoms will harmonize with the nearby English lavender and the violet gradations of the viola yesterday, today, and tomorrow, sown from the seeds that just arrived. In my imagination I can see how all the violet shades will complement the bright nasturtiums and pansies.

### Color in Medieval Tapestries and Manuscripts

*People in the Middle Ages loved bright primary colors. Their tapestries were filled with ruby reds, golds, cobalt blues, and grass greens. Their illuminated manuscripts were created by monks, who made many of their colors from plants in their cloister gardens. Blue came from the juice of elderberries or mulberries, purple from lichen, green from a mixture of iris and alum, and yellow from saffron crocus.*[10]

### Garden Exercise: Working with Color in Your Garden Design

AS YOU REVIEW your garden design and plan for the year ahead, consider how you can use color to further enhance your garden.

* *What colors predominate in your garden now?* Would you like to enhance your plantings by adding harmonious or complementary color?
* *Does one color tie your garden together?* Throughout the year my garden is interwoven with shades of violet. In the front yard purple crocuses and grape hyacinths bloom each spring, followed by cineraria and irises in a range of violets and purples. Summer brings Blue Moon hybrid tea roses, lavender, blue-violet lobelia, lilac sweet alyssum, and my neighbor's society garlic. From autumn to winter purple primroses bloom beside our garden gate. All around our yard are spots of violet, from the rose of Sharon beside my study window and the salvia that glows throughout the summer like an amethyst gem in our backyard to the tiny red-violet blossoms of *Hardenbergia violacea* cascading down our north garden fence in early spring. Even though my garden beds are filled with other colors, violet harmony ties the composition together, and next year I'll expand upon it.
* *Would you like to emphasize a favorite color* throughout your garden? If so, find perennials in this color that bloom at different times. Then add some annuals for a continuous display of color. Many gardening books have color guides listing plants of a specific hue that bloom in different seasons.[12]

### The Color Spectrum

*The colors in the spectrum—red, orange, yellow, green, blue, indigo, and violet (or "Roy G. Biv")—are electromagnetic waves. Red has the longest wavelength and violet the shortest. Colors are only a small part of the vast electromagnetic spectrum, which includes the longer radio waves, radar, microwaves, and infrared waves at one end and the shorter ultraviolet waves, X rays, and gamma rays at the other.[11]*

• *Would you like to experiment with color* in your garden next year, creating more dramatic contrasts or planting a monochromatic bed to accent your garden design?

Just asking yourself these questions and becoming more aware of color will enrich your palette and multiply the possibilities for your garden.

## GARDEN REFLECTION

### *Cultivating Winter Warmth and Color*

ON COLD JANUARY days I'm grateful for the bright spots of color in my garden: the tiny red berries on the cotoneaster and heavenly bamboo. Around town I've seen some stunning displays: pyracantha bushes laden with clusters of red-orange berries, late-season fruit hanging like golden ornaments on the bare branches of persimmon trees.

We can brighten our days by becoming more mindful of the colors around us and by bringing more color into our lives. Some gardeners force amaryllis bulbs to bloom indoors. I prefer to leaf through garden catalogs, savoring the colorful photographs, planning for tomorrow. This time of year I'm also drawn to jewel colors: ruby red, emerald green, amethyst, topaz, and sapphire blue. Wearing a colorful scarf or reaching for my purple fleece jacket before heading outdoors adds warmth and color to my days. Indoors I enjoy my needlepoint tapestries of medieval herbs and flowers. What can you do this month to bring more color into your life's design?

One way to ward off winter's chill is a steaming bowl of soup. After all the extravagant holiday fare, January calls us to simplify our lives with easy and satisfying meals. Soups and pottages were essential to the medieval diet, and in 1580 Thomas Tusser reminded his readers:

> Good broth and good keeping doo much now and than
> Good diet with wisedome best comforteth man.[13]

During the winter Bob and I have soup for dinner at least twice a week. Even when I'm at work all day, our soup cooks for hours in the Crock-Pot and welcomes us home at the end of the day. Yesterday I worked at home and made my favorite minestrone, which simmered for hours in a big pot on the stove, filled with carrots, potatoes, cabbage, onions, and green vegetables I had on hand.

## Heartwarming Winter Soups

IF YOU'D LIKE to make your own winter soups, here are some tips to get started.

Homemade soups begin with broth. You can make a classic broth by simmering pieces of chicken, turkey, or beef with some sliced onion, celery, and a bay leaf in a large pot filled with about four quarts of water for three to four hours. Or make a good vegetable broth with sliced onions, celery, cabbage, and summer squash, simmering with a bay leaf and four quarts for water for two hours. (You can skip this step by purchasing cans of ready-made broth at the grocery store.)

In many soups you first make the broth, then add the vegetables. For chicken soup, place three chicken breasts and a bay leaf in a pot with four quarts of water and simmer for an hour. Then slice one half of a small onion, two stalks of celery, and four carrots, adding them to the soup. Allow to simmer for one to two more hours. Remove the chicken breasts (you can eat them separately) and enjoy the soup, adding precooked rice or pasta if you wish.

For a simple minestrone, begin by making a classic meat or vegetable broth with four quarts of water, letting it simmer for one to two hours. Or use an equivalent amount of ready-made broth. Then chop and add whatever fresh vegetables you have on hand: onions, cabbage, carrots, celery, potatoes, broccoli, and green beans. Simmer for another hour or more. During the last hour add canned, fresh, or dried tomatoes. Stir in three tablespoons of pesto. Add rice or precooked pasta, and serve with Parmesan cheese sprinkled on top.

You don't need to cook all day to enjoy homemade soup with your meals. You can simply open a can or two of broth, pour it into a pan, and add whatever vegetables you have available: cabbage, onions, tomatoes, carrots, leftover potatoes or green beans, frozen corn or peas—you choose. Sprinkle in some dried basil or thyme. Add a tablespoon or two of cooking sherry if you like. Bring to a boil, then simmer on low for at least thirty minutes or until the vegetables are done. By the time you get the rest of your meal prepared, you will have a heartwarming hot soup for your first course.[14]

I like bringing my garden into the kitchen, even in winter, seasoning my soups with dried herbs, oven-dried homegrown tomatoes, and a few spoonfuls of pesto from last summer's garden. For part of gardening is always enjoying our harvest, which occurs on many levels throughout the garden year.

# FOURTEEN

## February:
## Seeds of Another Spring

What though the Heaven be lowring now,
And look with a contracted brow?
We shall discover, by and by,
A Repurgation of the Skie:
And when those clouds away are driven,
Then will appear a cheerfull Heaven.

ROBERT HERRICK,
"Hope well and Have well:
or, Faire after Foule weather"
(1648)

EARLY FEBRUARY FINDS us still in the midst of winter. In Northern California this means rain for weeks at a time. The skies are dark. The rain comes down in sheets and keeps me from my garden. It was raining again this Saturday, when I'd planned to dig the north garden beds and plant some early peas.

One day last week when I returned from work, the skies cleared briefly. In this short respite from the rain, I enjoyed walking around the front yard as the last rays lit up the garden at sunset. Reaching down to pull weeds from the moist clay soil, I discovered signs of new growth: green foliage rising from early spring bulbs. There were tiny red buds on the maple tree, and a few paperwhite blossoms brightened the winter landscape. But this time of year the rain often comes down all day, leaving the heavy clay soil too wet to work. Seeds planted now would only rot or be washed away.

While it frustrates my gardening efforts for now, we need the rain. Winter precipitation in Northern California provides our summer water supply. I remember too well the summer droughts when I struggled to keep my garden alive with "gray water" carried from the kitchen sink.

Rainstorms are a vital part of nature's cycle. The nineteenth-century American writer and naturalist Henry David Thoreau recognized this long ago during the rainy season at Walden Pond:

> The gentle rain which waters my beans and keeps me in the house today is not drear and melancholy, but good for me too. Though it prevents my hoeing them, it is of far more worth than any hoeing. If it should continue so long as to cause the seeds to rot in the ground and destroy the potatoes in the low lands, it would still be good for the grass in the uplands, and being good for the grass, it would be good for me.[1]

From close companionship with nature, Thoreau learned the wisdom of patience and process, vital lessons for this transition time in our gardens and our lives.

February takes us from winter to the first stirrings of spring. The cold, dark days are gradually growing brighter, the earliest spring bulbs awakening from their long winter sleep. In the Middle Ages and the

*While rocking Winds are*
*Piping loud,*
*Or usher'd with a shower*
*still,*
*When the gust hath*
*blown his fill,*
*Ending on the rustling*
*Leaves,*
*With minute-drops from*
*off the Eaves.*

JOHN MILTON,
"Il Penseroso,"
ll. 126–30 (c. 1631)

*Why, what's the matter*
*That you have such a*
*February face,*
*So full of frost, of storm*
*and cloudiness.*

WILLIAM
SHAKESPEARE,
*Much Ado About*
*Nothing,*
V.4, ll. 40–42, (1598)

Renaissance, our ancestors celebrated February 2 as the feast of Candlemas, commemorating Mary's presentation of the infant Jesus at the temple, forty days after his birth. Affirming the return of light to the world, priests blessed candles on this day and gave them to their congregations. In even earlier times Romans carried torches to honor the goddess Ceres and her search through the darkness of winter for her lost daughter Proserpine, who returned each spring from a four months' sojourn in the underworld.[2]

Old English customs for this time of year prepared people for a season of renewal as they removed the withered Yuletide holly, rosemary, bay, and mistletoe from their homes and churches, replacing them with fresh boxwood boughs. As the liturgical cycle turned toward spring, Candlemas brightened the last days of winter with the promise of Easter.

Other feasts this month include February 3, the Feast of St. Blaise, when the priest would bless the throats of parishioners, holding two blessed candles under their chins.[3] But today most of us associate February with Valentine's Day, February 14. St. Valentine was a martyr who died in Rome around 270 C.E., but because Valentine's Day occurred on the eve of Lupercalia, the ancient Roman fertility feast, this day soon became associated with romantic love.

In the Middle Ages, St. Valentine's Day was known as a time when young men and women, birds and beasts sought out their mates.

Medieval and Renaissance lovers exchanged Valentine gifts and tokens; the first paper valentines date back to the sixteenth century. According to one legend, if on February 14 a young woman would chant and run around a church twelve times or sleep with bay leaves and rose water under her pillow, she would find her future husband in her dreams.[4]

*Down with the Rosemary
  and Bayes,
Down with the Misleto;
In stead of Holly, now up-
  raise
The greener Box (for
  show).*

*The Holly hitherto did
  sway;
Let Box now domineere;
Untill the dancing
  Easter-day
Or Easters Eve
  appeare. . . .*

*Thus times do shift; each
  thing his time do's hold;
New things succeed, as
  former things grow old.*
      ROBERT HERRICK,
      "Ceremonies for
      Candlemasse Eve,"
      ll. 1–8, 21–22 (1648)

## GARDEN GROWTH

*Oft have I heard both*
*  Youths and Virgins say,*
*Birds chuse their Mates,*
*  and couple too, this*
*  day:*
*But by their flight I never*
*  can divine,*
*When I shall couple with*
*  my Valentine.*
        ROBERT HERRICK,
        To his Valentine, on
        S. Valentine's Day
        ll. 1–4 (1648)

*Saint Valentine is past:*
*Begin these wood-birds*
*  but to couple now?*
        WILLIAM
        SHAKESPEARE,
        *A Midsummer Night's*
        *Dream,*
        IV.1, ll. 136–37 (c. 1595)

DESPITE THIS MONTH'S stormy weather, my garden is showing signs of another spring. The camellia bushes along the north garden walk have begun to bloom in shades of deep rose, shell pink, and peppermint white flecked with red, their colors echoed by azalea blossoms in the front courtyard. Camellias (*Camellia japonica*) have a long history. Closely related to the tea plant (*Camellia sinensis*), camellias originated in ancient China, where they were cultivated in Buddhist monasteries. They were imported to Japan in 700 C.E. and came to Europe in the early 1700s.[5]

The buds on the bare branches of the flowering quince (*Chaenomeles*) by the courtyard gate have opened in a mass of deep rose blooms. Hardenbergia vines cascade down the north garden walls in an abundance of tiny purple blossoms. All over the front yard tiny violets raise their heads above clusters of heart-shaped leaves. Although these perennials (*Viola odorata*) bloom for only a brief time, they have inspired many poets. Scarlet and gold nasturtiums bloom nearby, thriving in the rainy weather. Their bright colors stand out dramatically beneath the bare winter branches and gray skies.

Early this month the first bulbs began to bloom. A golden crocus (*Crocus ancyrensis*) surfaced along the edge of the rose garden. Then hyacinths began appearing in the friendship garden, iris garden, and herb garden, raising their spires of lilac and white blossoms and filling the air with the sweetness of spring. Native to Greece and Asia Minor, hyacinths (*Hyacinthus orientalis*) were cultivated in Holland in the sixteenth and seventeenth centuries and from there spread to the rest of Europe.[6]

Joining the hyacinths, some tiny yellow narcissus blossoms (*Narcissus asturiensis*) appeared in the iris garden. Then by February 6 we had our first daffodil of the season, a golden *Narcissus* King Alfred, blooming in the north garden, followed in the next few days by several more. Violet Dutch crocuses (*Crocus vernus*) came up in the rose and iris gardens, and the first grape hyacinths (*Muscari armeniacum*) raised their blue-violet clusters, with paperwhites (*Narcissus tazetta*) blooming around the front gardens.

On Valentine's Day, after a week of torrential rains, the showers ceased, clouds parted, and rays of sunlight shone down on the gardens, once again at peace. The rains had knocked down some of the taller daffodils, so I propped them up. As I did I noticed more flowers: pumpkin and gold nasturtiums, new Dutch crocuses in Easter egg shades of lilac, purple, yellow, and white, lifting their cuplike blossoms from the earth. Legions of grape hyacinths have raised their standards around the front gardens, along with more daffodils and paperwhites. Here and there another hyacinth perfumes the air. The saturated soil makes the weeds easy to uproot. Spring blooms and their new foliage glisten with raindrops as I walk around the garden exploring, greeted by a damp, woodsy scent like that of a deep forest.

As the month continues the rain beats down in gray sheets, followed by more short reprieves when I find new flowers. More crocuses, in pastel yellow, lilac, and white, are coming up in the herb garden; new hyacinths are appearing around the yard; more daffodils, white poet's daffodils with gold centers (*Narcissus poeticus*), are joining the golden King Alfreds. Around the birch trees the perennial candytuft (*Iberis sempervirens*) is filled with white snowflake blossoms, and springing up all over the yard are the bright yellow

*I know a bank whereon
the while thyme blows,
Where oxlips and the
nodding violet grows.*

WILLIAM
SHAKESPEARE,
*A Midsummer Night's
Dream,*
II.1, ll. 249–50 (c. 1595)

*A single violet transplant,
The strength, the color,
and the size,
All which before was poor
and scant,
Redoubles still and
multiplies.*

JOHN DONNE,
"The Ecstasy,"
ll. 37–40 (c. 1601)

blossoms of Cape oxalis (*Oxalis pes-caprae*). After a week of rain, even these prolific weeds have their charm.

## Gardening as Spiritual Practice

### Patience and Process

THIS TIME OF year often makes gardeners impatient as we wander through the last days of winter, longing to begin spring planting. I really shouldn't complain. My sister- and brother-in-law in Massachusetts are still shoveling snow. My garden is graced by early bulbs while theirs is fast asleep.

Yet as our technology accelerates the pace of daily life, it's easy to lose patience with process. I live in California's Silicon Valley, which works overtime to produce faster and more powerful computers. But even outside the computer industry, the race for efficiency carries into much of our lives. As a society we've become accustomed to going faster and faster. How many of us become impatient commuting to work, waiting for computers to boot up, or even cooking our food in the microwave?

Time in our gardens brings us back to our senses. Beyond the illusion of control, the dominance of ego, we live in a natural world, where seasons move in cycles, seeds germinate in their own time, and natural processes cannot be rushed.

My garden always reminds me when I need to slow down and pay attention. The other day I was out weeding in a break between rainstorms. Rushing my way through the clumps of oxalis and dandelions, I yanked up a tiny bulb along with the weeds. Slowing down, I put the bulb back into the ground, hoping it would still grow, realizing how spoiled I am by media and microwaves. Garden tasks can help us practice patience and presence.

### Personal Exercise: Mindful Weeding

AS BENEDICTINE AND Buddhist monks have known for centuries, any repetitive task—working in the garden, washing dishes, chopping wood, carrying water—can become a meditative practice. Taking us away from today's frantic pace and impatient demands, a mindful practice reconnects us to the natural rhythms within and around us, opening us to a deeper source of insight and peace. The next time you're pulling weeds in your garden, try doing it more mindfully.

- *Don't just stand there staring at all you have to do.* Any task seems overwhelming when you consider it all at once.
- *Approach the task one small area at a time.* Look deeply at what you're doing.
- *Focus on the present moment.* What is your garden telling you? Notice any signs of pests or disease as well as small patterns of order and beauty.
- *You'll be rewarded* with glimpses of newly sprouted annuals and perennials. Hidden beneath the oxalis and dandelions, I found California poppy seedlings and clusters of reseeded sweet alyssum with purple and white blossoms as delicate as petit point. Nearby, bright purple crocuses were lifting their heads above the soil. These are the rewards of mindfulness.
- *There are also important messages.* What do the plants in your garden need? Is it time to feed bonemeal to emerging bulbs? Do they need protection from snails and slugs?
- *Gardening is an ongoing relationship.* Mindful weeding helps us get to know our gardens as we take time to watch, listen, take note, and enjoy the subtle patterns we would otherwise miss.
- *Any mindfulness practice carries over into other areas,* slowing us down to listen to ourselves, to notice the people and patterns around us, to cultivate important relationships at a depth impossible when we live by the frantic rhythms of our technology.

If it's too early to work in your garden, find another mindfulness practice. This year I decided to refinish an old chest of drawers I've had since graduate school. The chest is three feet high and three feet wide, with five drawers. When my aunt Betty gave it to me, it was covered with black lacquer. Through the years I've had it in my living room, bedroom, and study. And through my many moves it had become increasingly battered and scratched. I tried to conceal the scratches with black Magic Marker, promising myself that someday I'd refinish it.

When I moved into my new house in Los Gatos, shafts of sunlight from my study window exposed all the nicks and scratches on the old chest of drawers. After we got settled I took it out to the garage, spread a drop cloth and layer of newspaper, and bought some fine-grade steel wool and old furniture refinisher. I'd treated other furniture with this solution, which melts and renews the old finish with minimal effort. I expected the same result this time, figuring the job would take two or three days.

But this time was different. The black lacquer finish was thick, and sticky as molasses. Instead of spreading easily over the cracks, it adhered to the wood in a gloppy mass. When I'd worked on one section, the result looked worse than before: wood partially uncovered, smeared with tarry, molasses-like paint. When I finally removed the finish from one side of the chest, the wood itself was disappointing: coarse and inferior, with unmatched grains and a texture unlike that of any walnut antiques I'd refinished before. I went back to the hardware store for more cans of solvent. The stuff smelled horrible. Days, then weeks went by as I felt like Hercules cleansing the Augean stables. With no apparent end to my labors, I spent untold hours applying the foul-smelling solvent, wearing rubber gloves, a mask, and goggles.

Curious, Bob came out to the garage. Shaking his head, he inspected the hard-won result of my labors. "It looks like plywood," he said. "Why are you doing this? Wouldn't it just be easier to buy another chest of drawers?" His question made sense. For a little money I could buy a new piece of furniture, without any scratches or flaws, and throw the old one out. No matter what I did, this would still be an old, imperfect piece with nicks, scratches, and inferior wood.

Then I realized the issue was not perfection, economy, or convenience. It wasn't a new chest of drawers I wanted at all. I wanted to renew this one. Somehow this piece of furniture had become symbolic. It had come with me through all the stages of my adult life: graduate school, relationships, personal loss, moves, and repeated challenges. It had stood in my study while I'd finished my dissertation, prepared my first lectures, written my first books and articles. Now that I'd finally found the man and home I loved, I wanted to heal its battered surface and restore its beauty. But as the weeks went by I seemed to be making little progress.

One day I went to see my friend Chris, who showed me an old rocking chair that had been painted, neglected, and left outdoors before he rescued it. Years ago he had lovingly removed the paint and refinished it. It was beautiful, a venerable antique with a warm patina that wore its years like a legend. Chris had restored the chair by hand, using a piece of glass to take the paint out of the carved back, a process that took months. But he was in no hurry. He had the patience not to rush the process.

After hearing Chris's story I realized that some items take longer to refinish than others. That is their process. The sticky black finish on my chest of drawers would come off. It would just take time. When I told Chris the wood looked inferior and had no color, he assured me that a little Minwax would do the trick. "Just rub it in," he said.

I returned to my project with renewed hope. No longer rushing, I spent about an hour each day on the chest, one section, one drawer at a time, going over them again and again, as long as it took, to remove the old black molasses and get down to the natural wood. Telling myself not to worry about the way it looked, I rubbed steel wool and refinishing liquid over the wood, which gradually became cleaner and clearer. The wood grain didn't match, but that was no longer my concern. I simply did my part in the process.

Weeks later I got some Minwax and began rubbing it in, giving the wood new color and richness. One coat, then another, as long as it took. The days went by as I completed the Minwax applications and applied the last coat of tung oil varnish. I left the chest in the garage for a few days to dry, then brought it inside.

Today it sits in my study. Its once battered and chipped black finish is now a warm, mellow brown, the irregular wood grain and even the years of nicks and scratches blending into their own design. The chest of drawers is weathered, imperfect, *and* it is beautiful. Glowing with age and experience, it is much more to me than some new mass-produced piece of furniture.

This experience reminded me of the difference between beauty and perfection. When I finally put the drawers back in place, the pieces came together in their own natural pattern, the lines in the wood grain gradually narrowing as they flowed from left to right. When I'd been working with the individual parts, I could see only imperfections. Seeing the larger pattern, I discovered beauty. So it is with a garden and the gardens of our lives. There are blossoms and there are weeds, successful harvests and plants that never make it, moments of joy and disappointment woven into life's unfolding pattern. We can look for imperfections and we will find them. Perfection is static, and we live in an imperfect, ever-evolving world. When we stop demanding perfection, we can look beyond the details and discover the beauty of the larger design. Like our gardens, we are part of a larger process, a pattern of divine artistry that creates new harmonies within and around us.

## Questions to Consider

IF YOU'VE BEEN involved lately with a long process—restoring a piece of furniture, painting your house, repairing or building something—ask yourself these questions:

- What did I learn from this project? What did it teach me
  - about myself?
  - about patience, faith, and perseverance?
  - about the universe?
- What lessons can I apply in the future as I cultivate my garden and my life?

## Planting Seeds of Hope

LAST THURSDAY I came home after a long day of classes, conferences, meetings, and campus politics, discovering once again the restorative power of gardening. Gardens for me are like the "green world" in Shakespeare's comedies, where people are healed and taught important lessons, experiencing the transforming powers of nature. No matter what else happens, the simple acts of touching the earth diminish the unimportant problems of life, returning us to ourselves. The word *paradise* once meant an enclosed garden, and that meaning still holds true. In our gardens we touch paradise, regaining what our busy world has lost.

Today, since it's still too rainy to work outside, I tried some indoor gardening, sowing herbs in peat pots in my greenhouse box. A few days earlier I'd put some mung beans in a sprouting tray in the kitchen. This afternoon I noticed that the beans had already begun to sprout, soon to provide fresh greens for our salads. No matter how often I plant seeds, I always consider their sprouting a small miracle and the first tiny green seed leaves an affirmation.

*. . . in this pleasant soil*
*His far more pleasant*
*Garden God ordain'd;*
*Out of the fertile ground*
*he caus'd to grow*
*All Trees of noblest kind*
*for sight, smell, taste;*
*And all amid them stood*
*the Tree of Life.*
JOHN MILTON,
*Paradise Lost,*
IV, ll. 214–18 (1674)

## Personal Exercise: Planting Seeds

YOU, TOO, CAN plant seeds, both physically and spiritually, in this transition from winter to spring.

*Plant seeds inside* in a greenhouse box or a small tray of potting soil covered with plastic wrap. Keep the seeds moist and put them where they will stay warm enough for germination (check the seed packet, for these temperatures vary). When the seedlings emerge, remove the cover and set them on a sunny windowsill or under a growth light.[7] Feed them each week with a mild fertilizer, following the directions on the package. In six to eight weeks your plants should be ready to harden off. As weather permits leave them outdoors for a few hours at a time, gradually increasing the time to overnight. Then plant them in your garden.[8]

*Plant the seeds of tomorrow in your life.*

- Look over your dream/action card and your life design sheet from last month.
- How are you doing? Have you persevered and made progress? Great!

How can you keep moving forward? Write this down on a new action card for the month ahead.

✐ If you haven't progressed as you wished, do you need to revise your plan or recommit? Then plant the seeds of determination. Write down the action you've chosen on a new index card, sign and date it.

✐ Put your action card where you'll see it at least once a day.

✐ Mark on your calendar when you'll take action, beginning tomorrow.

✐ Smile as you see yourself approaching your goal.

*Have you discovered a new direction to your dream?*

✐ If so, write this new "seed" habit or project on an index card and put it where you'll see it at least once a day.

✐ Make a list of steps. What will you need to do to get to your goal?

✐ Number the items on the list and mark your calendar to take one small step in this new direction tomorrow. Then check this item off your list with a smile.

*Keep planning and taking action.*

✐ Each day when you look at your action card, take a moment to cultivate your dream by visualizing what the result will look and feel like.

✐ Smile, take a deep breath, and go on with your day.

Now that you've planted the seed and cultivated it, you'll be surprised at all the new opportunities you'll discover.

Whether the seeds you plant are physical or psychological, you'll be working with the powers of the universe to make things grow. In his twenty-seven years of imprisonment, Nelson Mandela found hope in two projects: writing his life story and gardening. "To survive in prison, one must develop ways to take satisfaction in one's daily life," he said. During his time on Robben Island, he found ways to write his manuscript in secret, hiding it and eventually smuggling it out. Then there was his garden. After working all day in a quarry, he'd spend much of his free time digging in the hard, rocky soil of the prison courtyard. There he grew tomatoes, chilies, and onions, watching them grow, nurturing them, sharing his harvest. Later, on the roof of his prison at Pollsmoor, he grew more vegetables: beans, spinach, eggplants, cabbage, broccoli, beets, lettuce, and strawberries as well as tomatoes, onions, and peppers. His own words about gardening reveal his indomitable spirit:

> A garden was one of the few things in prison that one could control. To plant a seed, watch it grow, to tend it and then harvest it, offered a simple but enduring satisfaction. The sense of being the custodian of this small patch of earth offered a small taste of freedom.[9]

Nelson Mandela experienced the spiritual growth all gardeners feel when they work with nature's process. In his garden he planted seeds of hope that grew into unprecedented patterns of renewal for himself and the new South Africa.

## GARDEN TASKS

### Early Season Care

IN THE MIDDLE Ages people returned to their gardens in February for early cultivation and planting. When they finished their pruning, they began turning over the earth. In warmer regions they planted early crops of oats, barley, and peas.[10] For all of us the garden year begins and ends at different times. Some regions are still covered with snow, while in others the earliest bulbs are raising their heads to welcome us back to the garden.

In my Northern California garden, it's been raining all week, but this morning during a break I went out to tend the garden. Spinach and lettuce seedlings, planted last week, are coming up in planters, so I fed them with fish emulsion. I sprinkled snail bait all around because I've seen a few snails and spotted some damage. I deadheaded spent narcissus and daffodils and trimmed back the ferns and shrubs growing along our north garden walk.

The rain has knocked over many blooming daffodils, but other bulbs are raising their foliage, and I spied some new violet and yellow crocuses along the edge of the rose garden. I started pulling up oxalis along the north garden walk, but then it began to rain again, so I came back inside, noticing the raindrops clinging like pearls to the bare branches of the Japanese maple.

*And now in age I bud*
  *again,*
*After so many deaths I*
  *live and write;*
*I once more smell the dew*
  *and rain,*
*And relish versing: O my*
  *onely light*
*It cannot be*
*That I am he*
*On whom thy tempests*
  *fell all night.*
GEORGE HERBERT,
"The Flower,"
ll. 36–42 (1633)

In my area February care includes pruning deciduous trees and vines, deadheading spent bulbs, weeding, feeding early blooms and greens, setting out snail and slug bait, picking up fallen camellia blossoms, protecting tender plants from frosts, planting bare rootstock and early vegetables. But it's been such a wet February this year that most of my gardening has been limited to short periods of trimming, weeding, and tidying up.

The other day it had stopped raining when I came home from school, so I put down my briefcase and reached for my gardening gloves. I pulled up some oxalis from the herb garden and cleared out more oxalis and fallen leaves around the ajuga in the front courtyard so these plants can awaken from winter, spreading their plum-tinged foliage across the ground. A few days later, after working at home to meet a publication deadline, I went outside at five o'clock. The clouds had parted after a day of showers. The air was fresh, the ground moist, the cobblestone path shining, and new foliage from early bulbs bright spring green. As I pulled up more oxalis, giving the emerging irises room to grow, I took a deep breath, feeling a profound sense of peace.

## Early Spring Planting

THOMAS TUSSER'S GUIDE for Renaissance gardeners says that February is the time to dig deep trenches in our garden beds, fill them with compost and manure, and then sow early peas.[11] In February many of us can begin planting the earliest spring vegetables: sowing them in the soil in zones 7 through 11, covering them with cloches or hot caps where the weather is still cold. As soon as the ground can be worked, the earliest vegetables — lettuce, spinach and other greens, radishes, peas, and beets — can be planted, although in zones 4 through 8 you'll need to use a cold frame or sow your seeds indoors.

Each year you'll have to watch local conditions to know when to plant. The best time to sow and set out new plants depends on annual changes in the growing cycle (spring comes earlier some years, later in others) and weather conditions (if it's too wet, your seeds will not germinate).

Early this month I planted some spinach and lettuce seeds in planters along our south garden fence. By the time I want to plant tomatoes in the same planters, I'll have harvested these early greens. Feeling industrious one day, I bought three bags of potting soil and a bag of organic compost at the hardware store. Between rain showers I put compost and potting soil down to make a raised bed in the south garden, then set a two-by-four board there to hold in the soil. In the deep, loose soil of this raised bed, I sowed my seed potatoes. Potatoes can be planted in a sunny bed with deep, loose soil two or three weeks before the last frost date. When the potato plants appear, I'll add mulch around them to provide more room for the potatoes to grow.

I dragged the two other bags of potting soil out to the north garden beds, where I dug them in, adding some bonemeal and turning over the soil, which looked impressively rich and loose textured, ready for the peas. On February 7 I sowed my first crop of peas, after treating them with an inoculant to increase the nitrogen. I was so eager to get them into the ground that I ended up planting them by moonlight. Time will tell if my sowing was too early or if I will harvest these "moonlight peas."

A few days later the blueberry bushes and bay tree I'd ordered arrived in the mail. I carefully unpacked the tiny plants from the shipping box, removed their straw insulation, and checked the soil in their pots. It was still moist, so I misted their leaves and set them outside in the front courtyard until I had time to plant them.

I planted the blueberries in twin redwood planters, adding azalea mix to the soil because they like their soil acidic. The blueberry bushes are only eight inches tall. Who knows how long it will take, but someday we'll have fresh blueberries to harvest in our own front yard.

My bay laurel tree is shorter than the nearby bonsai. In their native Mediterranean countries, laurel trees grow forty feet tall. I planted mine in a redwood planter so I can bring it beside the house to protect it from frost. Eventually I'll have to prune it so it won't outgrow its container. But all that is well into the future. For now it's still small and its leaves too precious to harvest for kitchen use.

I've always wanted to grow a laurel tree, not only for its aromatic and flavorful leaves but for its classical associations. For the ancient Romans the laurel was sacred to the god Apollo. According to myth, the nymph Daphne escaped from Apollo's amorous pursuit by becoming a laurel tree. I remember translating that myth from Ovid's *Metamorphoses* in my ninth-grade Latin class, so surprised by the transformation I wondered if I'd gotten the words wrong. Apollo claimed the laurel as his emblem, and its branches became symbols of victory, honoring conquering heroes and poets. Laurel wreaths were used in medieval universities to crown recipients of academic degrees in poetry, grammar, or rhetoric. Petrarch was crowned the first poet laureate in Rome in 1341, beginning a tradition that continues in many countries. In medieval England the laurel was also an emblem of constancy.[12]

## Planting Seeds of Spring

IN FEBRUARY MANY of us begin planting seeds for our spring gardens. I can sow beets, carrots, lettuce, spinach, and peas, as well as hardy annuals. But even if it's still winter outside, you can cultivate another springtime in your heart by sowing seeds indoors.

Many annuals and vegetables can be grown indoors now. To decide which ones are appropriate for your region, a little math is in order. Determine when you'll want to set the young plants outside, then count back six to eight weeks. If by mid-April spring weather usually comes to your region, then you can start some seedlings indoors now. You'll want to start the cool weather flowers and vegetables first, referring to the seed packets for specific information.

## *Growing Healthy Seedlings*

THERE'S SOMETHING WONDERFUL about starting seeds of spring when the world outside still feels like the dead of winter. The tiny seeds germinate, lifting their green leaves as signs of hope. To help your seedlings stay healthy, here are some common afflictions and treatments:

- Long, weak, leggy stems. Cause: Inadequate light, overcrowding, or excessive heat. Treatment: Move the plants to give them more light and a cooler temperature; thin them to give them more room.
- Seedlings wilting and bending over. Cause: Insufficient moisture; the plants are drying out. Treatment: Soak the soil and make sure it stays moist.
- Mold or fungus on the soil. Cause: Excessive moisture, insufficient drainage. Treatment: Give the plants more air; remove the cover, cut back on watering, move to a spot with more air circulation.
- Seedlings falling over and collapsing at the base. Cause: Damping off, a fungus infection. Treatment: Remove afflicted seedlings; move the container to a place with more air, light, and drainage; spray the seedlings with chamomile tea, which helps fight infection.
- Pale leaves. Cause: insufficient nitrogen. Treatment: Feed with liquid fertilizer. If you grow your own sprouts for salads and stir-fry dishes, save the rinse water. It's rich in gibberellins, growth hormones, which give an added boost to seedlings.

Even if you don't want to start seedlings inside, you can still cultivate spring by sprouting seeds in your kitchen for salads and stir-fry dishes. On dark, stormy days when you can't get out to the garden, you can harvest fresh green sprouts. I've grown sprouts for years in an old quart jar with a piece of screen or muslin on top. This year I got a clear Plexiglas sprouting

**Growing Your Own Sprouts**

*To grow your own sprouts:*

- *Take an old canning jar, fill it with 2 tablespoons or so of seeds (such as organic lentils or mung beans from the health food store).*

- *Fill the jar with water, and cover the top with a piece of cheesecloth or muslin held in place by a rubber band.*

- *Let the seeds soak for a while (a few hours to overnight). Then pour out the water and lay the jar on its side.*

- *Rinse the seeds with water once a day.*

- *In a few days you'll have fresh, new sprouts for stir-fries or salads. You'll also witness the small miracle of seed germination right before your eyes.*

container with three levels. This little sprout "greenhouse" on my kitchen counter gives me a view of nature in action as the tiny sprouts germinate and grow right before my eyes. There was a frost warning last night, but indoors the mung and azuki beans I'd put in the sprouter on Monday were ready to harvest for Friday's dinner.[13]

## GARDEN REFLECTION

### *Cultivating Thyme*

I'VE ALWAYS LIKED to cook with thyme: thyme sprinkled on oven-roasted new potatoes drizzled with olive oil; chicken breasts roasted with olive oil, honey, and thyme; thyme sprinkled in soups; Dover sole poached in white wine and thyme. I picked some thyme from the herb garden and dried it in the oven this fall, storing it in a tea canister. This winter it flavored many baked and roasted dishes, but all that thyme is gone. The two English thyme plants in the garden have been stunted by the cold. What I need now is more thyme: thyme to season a myriad of simple recipes, dried thyme to flavor my winter meals, fresh thyme to sprinkle on summer homegrown tomatoes.

This year I want more thyme in the garden, creeping thyme to grace the cobblestone walk with fragrance, thyme to slow my daily comings and goings so I discover new blossoms and pause to enjoy the subtle miracles of nature. More thyme in the garden to sweeten the time of my life.

A few weeks ago I ordered a packet of creeping

thyme (*Thymus serpyllum*) to sow between the stepping-stones, among the clover and grasses. I'd thought of simply sowing it in situ, letting the rains this month help the seeds germinate. But the rains have been torrential and the tiny seedlings would drown in the muddy puddles. If I want more thyme, I'll have to plant my seeds more carefully. Cultivating thyme takes attention to detail, mindful choices.

I've just finished a three weeks' blitz of work with one press deadline after another, along with midterm exams and endless meetings at school. Among other things I'm faculty senate president this year. The garden of my life has become crowded with too many activities, and I long for more margins, fragrant borders, and spaces for contemplation.

Cultivating thyme, I reasoned, could be a glorious pun, a spiritual exercise and ongoing affirmation. Since I'm sowing the seeds indoors in my greenhouse tray, taking the time to water and check the seedlings will purposefully slow me down, making me more mindful of small things. Watching each day for signs of germination, celebrating the emerging seedlings, setting them under the growth light, and, in good time, transplanting them outdoors will become an ongoing spiritual practice, a recurrent lesson in tending more wisely the garden of my life.

On February 24 I prepared the greenhouse box, dampened the peat pots, and carefully placed the tiny seeds into the soil. They were as small as grains of sand in an hourglass. The instructions said they'd take up to three weeks to germinate. On February 27 I could hardly believe my eyes. A few seed leaves, each no bigger than the head of a pin, had pushed their way out of the soil and were reaching for the light. The next day more seedlings appeared. I set the flat under the growth light and removed the lid, initiating the seedlings to their next stage of growth. I don't know how long it will take, and that's not my concern. In time these tiny seedlings will be ready to move outside.

When the new plants take hold and grow along the path to my front door, their fragrance will sweeten my comings and goings with a reminder to embrace the daily gift of time. For that's what our gardens do for us, really. They remind us that beneath all the activities in our lives that assume such chronic urgency, there are much larger and richer patterns within and around us. We are part of nature's ongoing cycle. Beyond the frantic rush of schedules and digital time is the seasonal rhythm of beginnings and endings, that green emergence of life from darkness, the miracle of life we witness each day in our gardens.

At once humbling and empowering, wide-reaching yet deeply personal, gardening teaches us lessons in patience, hope, and fortitude, affirming values that transcend all human measure. In patterns eternal yet forever new, the lives of herbs and flowers blend with our own in the process of inner gardening, bringing us perennial gifts of grace written in the language of the heart.

May your life be blessed with joy and renewal through all your garden years.

# $\mathcal{N}otes$

*Chapter One: Gardens and Personal Growth*

1. Edward Hyams, *A History of Gardens and Gardening.* (New York: Praeger, 1971), pp. 19, 64.

2. Teresa McLean, *Medieval English Gardens* (New York: Viking, 1980), p. 14.

3. Hyams, *History of Gardens,* pp. 95–97.

4. Thomas More, quoted in Eleanour Sinclair Rohde, *Herbs and Herb Gardening* (New York: Macmillan, 1937), p. 18.

5. Andrew Marvell, "The Garden," ll. 7–8, in *Seventeenth-Century Prose and Poetry,* 2d ed., ed. Alexander Witherspoon and Frank J. Warnke. (New York: Harcourt Brace, 1982), p. 969.

6. E. J. Langer and J. Rodin, "Effects of Choice and Enhanced Personal Responsibility for the Aged: A Field Experiment in an Institutional Setting," *Journal of Personality and Social Psychology,* 34 (1976), 191–99. In this study, one practice that helped nursing home residents live longer, healthier lives was choosing a plant and caring for it.

Quotations

François Marie Arouet de Voltaire, *Candide ou L'Optimisme* (1759); Paris: Bordas, 1969), p. 185. The final line of the book is *"Cela est bien dit, répondit Candide, mais il faut cultiver notre jardin."* I am grateful to Catherine Montfort, Professor of Modern Languages and Literatures, and Helene LaFrance, Senior Assistant Librarian, Orradre Library, Santa Clara University, for their assistance in locating the quote. The English line cited is my adaptation from the French.

*Chapter Two: Gardens Past and Present*

1. Marvell, "The Garden," ll. 47–48.

2. See Sylvia Landsberg, *The Medieval Garden* (London: British Museum Press, n.d.), p. 56; Umberto Eco, *Art and Beauty in the Middle Ages*, trans. Hugh Bredin (New Haven: Yale Univ. Press, 1986), p. 46.

3. See Langer and Rodin, "Effects of Choice," and the discussion of remarkable older women in Peggy Downes, et al., *The New Older Woman* (Berkeley, Calif.: Celestial Arts, 1996), p. 11.

4. See "paradise" in *The Compact Edition of the Oxford English Dictionary* (New York: Oxford Univ. Press, 1971), p. 2072 (p. 449 of the original edition). See also Christopher Thacher, *The History of Gardens* (Berkeley: University of California Press, 1979), p. 15; McLean, *Medieval English Gardens*, pp. 16, 18, and 31; and Howard Loxton, ed., *The Garden* (Toronto: Key Porter, 1991), p. 22.

5. McLean, *Medieval English Gardens*, p. 16; Elizabeth and Reginald Peplow, *In a Monastery Garden* (London: David & Charles, 1988), pp. 8, 26.

6. McLean, *Medieval English Gardens*, pp. 15, 20, 41; Peplow and Peplow, *In a Monastery Garden*, p. 50–51; Hyams, *History of Gardens*, p. 90.

7. Bonaventura's advice to meditate on the Book of Nature is found in *Itinerarium Mentis ad Deum*. A good English translation is Bonaventura, *The Mind's Road to God*, trans. George Boas (Upper Saddle River, N.J.: Prentice-Hall, 1953). For St. Ignatius's vision of God's presence in the natural world, see the "Contemplation to Attain the Love of God" at the end of *The Spiritual Exercises of St. Ignatius*. trans. Louis J. Puhl, S. J. (Chicago: Loyola Press, 1951), pp. 102–3. I am grateful to William Rewak, S. J., for this insight into Ignatian spirituality.

8. See Thacher, *History of Gardens*, p. 83; Landsberg, *Medieval Garden*, p. 53; and Hyams, *History of Gardens*, p. 93.

9. See Candace Bahouth, *Flowers, Birds, and Unicorns: Medieval Needlepoint* (London: Conran Octopus, 1993), pp. 51–53.

10. Eco, *Art and Beauty*, p. 45; Bahouth, *Flowers, Birds, and Unicorns*, p. 51; Jon Gardener, *The Feate of Gardening* (c. 1400) cited in Hyams, *History of Gardens*, p. 95; Landsberg, *Medieval Garden*, pp. 83, 52–53; McLean, *Medieval English Gardens*, pp. 140, 142, 169, 184, 186, 194.

11. Landsberg, *Medieval Garden*, pp. 52–53.

12. McLean, *Medieval English Gardens*, pp. 160, 139.

13. Loxton, *The Garden*, p. 25.

14. Peplow and Peplow, *In a Monastery Garden*, p. 52; McLean, *Medieval English Gardens*, p. 80.

15. Thacher, *History of Gardens*, pp. 126–27.

16. *Virgil's Georgics* (1697), II, l. 655, trans. John Dryden, ed. Alan Roper and Vinton A. Dearing, in *The Works of John Dryden*, ed. H.T. Swedenberg vol. 5 (Berkeley and Los Angeles: Univ. of California Press, 1987), p. 202.

17. John Milton, *Paradise Lost*, XII, l. 587 in *John Milton: Complete Poems and Major Prose*, ed. Merritt Y. Hughes (New York: Odyssey Press, 1957), p. 467.

18. Milton, "Of Education," pp. 634–35.

19. Pope, quote from letter to Lady Mordaunt, 1736, quoted in Hyams, *History of Gardens*, p. 236. See also Peter Martin, *Pursuing Innocent Pleasures: The Gardening World of Alexander Pope* (Hamden, Conn.: Archon, 1984), pp. xix, xxi., 10–11, 19, 21.

20. Jefferson, letter to Charles Willson Peale (August 20, 1811), quoted in Loxton, *The Garden*, pp. 78–79. For his lifelong garden observations, see *Thomas Jefferson's Garden Book*, ed. Edwin Morris Betts, introd. Peter J. Hatch (Monticello, VA: Thomas Jefferson Memorial Foundation, 1999).

## Quotations

Marvell, "The Garden," ll. 71–72, *Seventeenth-Century Prose and Poetry*, p. 970.

Joseph Addison, *The Spectator*, no. 477, September 6, 1712.

Addison, *Spectator*, no. 477.

Anne Collins, "Song," from *Divine Songs and Meditacions* (1653), p. 57, cited in Stanley Stewart, *The Enclosed Garden* (Madison: Univ. of Wisconsin Press, 1966), p. 108.

Christopher Harvey, *The School of the Heart* (1664), p. 125 quoted in Stewart, *Enclosed Garden*, p. 51.

Abraham Cowley, "The Wish," ll. 10–12, *Seventeenth-Century Prose and Poetry*, p. 954.

Marvell, "The Garden," ll. 65–72, p. 970. For information on sundial gardens, see Stewart, *Enclosed Garden*, pp. 100–01.

Dryden, *Virgil's Georgics*, II, ll. 698–701, p. 203.

*Chapter 3: March*

1. Dorothy Gladys Spicer, *Yearbook of English Festivals* (New York: H. W. Wilson, 1954), pp. 39, 41; Marie Collins and Virginia Davis, *A Medieval Book of Seasons* (New York: HarperCollins, 1992), pp. 43–44; R. Chris Hassel, *Renaissance Drama and the English Church Year* (Lincoln: Univ. of Nebraska Press, 1979), p. 112.

2. Stewart, *Enclosed Garden*, p. 123.

3. See, for example, Elizabeth Gould, et al., "Neurogenesis in the in the Neocortex of Adult Primates," *Science*, 286 (October 15, 1999), pp. 548–52.

4. A soil test kit will help you determine any necessary additional nutrients. Simple kits that test PH, nitrogen, phosphorus, and potassium levels are available at hardware and garden supply stores. In *Southern California Gardening* (San Francisco: Chronicle Books, 2000), p. 19, Pat Welsh recommends adding gypsum to some clay soil to improve drainage.

5. For information on double-digging, I am grateful to John Jeavons for his classic, *How to Grow More Vegetables Than You Ever Thought Possible on Less Land Than You Can Imagine* (Berkeley, Calif.: Ten Speed Press, © 1995 Ecology Action), pp. 7–14. Information used by permission. For more advice about double-digging, soil testing, composting, and organic gardening, contact the Common Ground Store at 2225 El Camino Real, Palo Alto, CA 94306, (650) 328–6752, www. commongroundinpaloalto.org or Bountiful Gardens, 18001 Shafer Ranch Road, Willits, CA 95490-9626 (707) 459-6410. On-line catalog and organic gardening product ordering information also available at www.bountifulgardens.org. For workshops and information about Ecology Action's worldwide "Grew Bio intensive" sustainable mini-farming work, see www.growbiointensive.org.

6. McLean, *Medieval English Gardens*, p. 219.

7. Tusser, *Five Hundred Points*, intro. Geoffrey Grigson (Oxford: Oxford Univ. Press., 1984) p. 38. Used by permission.

8. McLean, *Medieval English Gardens*, pp. 105, 199.

9. A good resource for information on composting is Stu Campbell's book, *Let It Rot: The Home Gardener's Guide to Composting* (Pownal, Vt.: Storey Communications, 1975). Jeavons also has a helpful composting chapter (pp. 30–42).

10. Tusser, *Five Hundred Points*, p. 94.

11. Inge N. Dobelis et al. eds, *Magic and Medicine of Plants* (Pleasantville, N.Y.: Reader's Digest, 1986), p. 159; Susun S. Weed, *Menopausal Years: The Wise Woman Way* (Woodstock, N.Y.: Ash Tree, 1992), pp. 46, 64.

12. Tusser, *Five Hundred Points*, p. 88.

13. Hyams, *History of Gardens*, p. 2; Dobelis et al., *Magic and Medicine*, p. 261; Diane Dreher, *The Tao of Womanhood* (New York: William Morrow, 1998), p. 45. For excellent advice on growing onions, see Welsh, *Southern California Gardening*, pp. 284–86.

Quotations

William Shakespeare, *The Winter's Tale*, IV.4, ll. 118–20, in *The Norton Shakespeare*, ed. Stephen Greenblatt et al. (New York: W. W. Norton, 1997), p. 2924. Text © 1988, Oxford Univ. Press. Used by permission.

Dryden, *Virgil's Georgics*, I, ll. 463–64, p. 171.

Bonaventura, "Sermon on St. Francis," (Paris, October 4, 1255), in *The Disciple and the Master: St. Bonaventura's Sermons on St. Francis of Assisi*. trans. and ed. Eric Doyle, O.F.M. (Chicago: Franciscan Herald Press, 1983), p. 61. Used by permission of the Franciscan Press, Quincy University, Quincy, Ill. 62301.

Dryden, *Virgil's Georgics*, I, ll. 64–69, p. 157.

Ibid., II, ll. 87–88, p. 183.

Tusser, *Five Hundred Points*, p. 71.

*Chapter 4: April*

1. Milton, *Paradise Lost*,VIII, ll. 352–53, p. 371.

2. Shakespeare, *A Midsummer Night's Dream*, V.1, ll. 16–17, p. 851.

3. For more information on botanical names, see William T. Stearn, *Botanical Latin*. 4th ed. (Portland, Oreg.: Timber Press, 1992). *The Garden Gate* Web site also has some helpful explanations of botanical names at www.prairienet.org/ag/garden/botrts.htm.

4. For information on Easter and traditional April feasts, see Ethel L. Urlin, *Festivals, Holy Days, and Saints' Days* (1915; Ann Arbor: Gryphon Books, 1971), pp. 58, 73, 80, 93; Spicer, *Yearbook*, pp. 44, 189.

5. Dobelis, et al., *Magic and Medicine*, p. 128.

6. Hyams, *History of Gardens*, pp. 4–5.

7. Dobelis, et al., *Magic and Medicine*, p. 352; McLean, *Medieval English Gardens*, p. 143.

8. For information and recipes for nasturtiums and other edible flowers, see Renee Shepherd's Kitchen Garden column in the magazine section of www.Garden.

com. But be careful when you add flowers to your cuisine. Not all flowers are edible. Daffodils and foxgloves are highly toxic, and any flower (including nasturtiums and roses) sprayed with pesticides should not be eaten.

9. Dobelis, et al., *Magic and Medicine*, pp. 233, 285; McLean, *Medieval English Gardens*, p. 184; Nicholas Culpeper, *Culpeper's Complete Herbal and English Physician Enlarged* (1814); Glenwood, Ill.: Meyerbooks, 1990), pp. 155–56.

10. Kathleen Brenzel et al., eds. *The Sunset Western Garden Book*. (Menlo Park, Calif.: Sunset, 1995), p. 351.

11. 1 Corinthians 12:4 in *King James Bible* (1611).

12. For more information on environmentally friendly pest control, consult the University of California Integrated Pest Management (IPM) Web site: www.ipm. ucdavis.edu, see Helen and John Philbrick, *The Bug Book* (Pownal, Vt.: Storey Communications, 1974), or visit the Gardens Alive Web site: www.gardens-alive. com, which has a helpful section on pest control.

13. For more information on the heat-zone map, contact the American Horticultural Society, 7931 East Boulevard Drive, Alexandria VA 22308, (800) 777-7931, or visit their Web site: www.ahs.org.

14. See *Thomas Jefferson's Garden Book*, pp. 87–89.

15. Thermometers that test soil and compost temperature are available by mail order from Bountiful Gardens, 180001 Shafer Ranch Road, Willits, CA 95490-9626, (707) 459-6410. Their on-line catalog and ordering information are on www.bountifulgardens.org.

16. Inoculants for beans and peas are available from many sources, including Bountiful Gardens, 180001 Shafer Ranch Road, Willits, CA 95490, (707) 459-6410, www.bountifulgardens.org. and Shepherd's Garden Seeds, 30 Irene Street, Torrington, CT 06790–6658 (860) 482-3638, www.shepherdseeds.com.

17. Peplow and Peplow, *In a Monastery Garden,* pp. 110, 114–116; McLean, *Medieval English Gardens*, p. 179; Culpeper, *Culpeper's Complete Herbal*, pp. 63, 82, 104, 113, 117, 131, 162, 183; John Gerard, *The Herball*, Rev. Thomas Johnson (1633; New York: Dover, 1975), pp. 178, 572, 584, 672, 766, 1014, 1033. 1293.

18. McLean, *Medieval English Gardens*, pp. 220–21; an article in *Prevention* magazine (May 1999), pp. 167–68, tells how honey may help heal minor cuts and prevent certain pollen allergies.

19. For an introduction to the Benedictine Divine Office and a discussion of the meaning of *Lectio*, see *Work of God: Benedictine Prayer*, ed. Judith Sutera, O.S.B. (Collegeville, Minn.: Liturgical Press, 1997), p. 99.

Quotations

The first two lines of Chaucer's *Canterbury Tales* in Middle English. A modern English version would be: "When April with his sweet showers has pierced the drought of March to the root . . ." From *The Works of Geoffrey Chaucer*, 2d ed., ed. F. N. Robinson (Boston: Houghton Mifflin, 1961), p. 17. Copyright © 1957 by Houghton Mifflin Company. Used with permission.

Milton, *Paradise Lost*, IV, ll. 694–703, p. 294.

Robert Herrick, "To Daffodills," ll. 1–2 from *Hesperides* (1648), in *The Poems of Robert Herrick*, ed. L. C. Martin (London: Oxford Univ. Press, © 1965), p. 125. Used by permission.

Herrick, "A Meditation for his Mistresse," ll. 1–3, p. 87.

Shakespeare, *Romeo and Juliet*, I.2. ll. 24–26, p. 879.

George Herbert, "Giddinesse," ll. 1–4, from *The Temple* (1633) in *The Works of George Herbert*, ed. F. E. Hutchinson (Oxford: Clarendon Press, 1945), p. 127 © 1941 Oxford University Press. Used by permission.

Dryden, *Virgil's Georgics*, II, ll. 386–89, p. 194.

Tusser, *Five Hundred Points*, p. 97.

Ibid., p. 87.

Ibid., p. 27.

Ibid., p. 94.

Ibid., p. 89.

## Chapter 5: May

1. For information on early May Day celebrations, see Urlin, *Festivals, Holy Days*, p. 105; Spicer, *Yearbook*, p. 57; and Nigel Pennick, *The Pagan Book of Days* (Rochester, Vt.: Destiny Books, 1992), p. 15.

2. For information on Christian May feasts, see Urlin, *Festivals, Holy Days*, pp. 99, 103, 118, and Spicer, *Yearbook*, pp. 229, 257, 267.

3. Chaucer, Prologue to *Legend of Good Women*, l. 42, p. 483.

4. For information about daisies in the Middle Ages, see McLean, *Medieval English Gardens*, pp. 117–118, 150, 161–62.

5. Information on thigmotropism from Peter H. Raven, Ray F. Evert, and Susan E. Eichhorn, *Biology of Plants*, 6th ed. (New York: W. H. Freeman, 1999), p. 706. I am grateful to William Eisinger, Professor of Biology at Santa Clara University, for sharing this informative book.

6. For historical information about roses, see Peter McHoy, *The Ultimate Rose Book* (New York: Hermes House, 1999), pp. 8–9; Loxton, *The Garden*, p. 22; and McLean, *Medieval English Gardens*, pp. 127–29.

7. For a discussion of roses in the medieval church, see McLean, *Medieval English Gardens*, pp. 31, 132, 168.

8. McHoy, *Ultimate Rose Book*, pp. 8–9. McHoy's book contains numerous rose recipes and crafts: potpourris, pomanders, sachets, oils, sugar, jelly, wine, cakes, and candies that echo the practices of medieval times. McLean, *Medieval English Gardens*, pp. 127–29; Loxton, *The Garden*, p. 22.

9. McLean, *Medieval English Gardens*, pp. 169–70.

10. History of roses drawn from McHoy, *Ultimate Rose Book*, p. 10. The story of the Peace rose from Derek Fell, *The Essential Gardener* (New York: Crescent Books, 1990), pp. 302–4. For more information about roses: descriptions of new and classic varieties and advice on cultivation, contact the American Rose Society, P.O. Box 30,000, Shreveport, LA 71130-0030, (800) 637-6534, e-mail ars@ars-hq.org, or visit the Rose Resource on-line at www.rose.org.

11. Shakespeare, sonnet 18, ll. 1–4, p. 1929. For information on medieval computation of seasons, see Collins and Davis, *Medieval Book of Seasons*, p. 28.

12. Bonaventura, *Mind's Road to God*, pp. 14–15. A discussion of the influence of Bonaventuran nature meditation on the poetry of the late Renaissance can be found in Louis L. Martz, *The Paradise Within* (New Haven: Yale Univ. Press, 1964).

13. Thomas Traherne, *Centuries of Meditation* (Wilton, Conn: Morehouse, 1986), pp. 13–14 (from meditations 27 and 29 of the first century). From *Centuries, Poems, and Thanksgivings*, ed. H.M. Margoliouth (Oxford Univ. Press © 1958).

14. For an excellent description of the biochemistry of photosynthesis, see Raven et al., *Biology of Plants*, pp.126–51.

15. My thanks to the Master Gardeners of Santa Clara County and Millie Wright, who shared this rose spray recipe at a gardening class at Campbell Community Center on April 12, 2000. The Master Gardeners, an outreach of the University of California, promote integrated pest management and remedies that minimize risk to ourselves, our soil, and our environment. For more information about Master Gardeners, call (408) 299-2638, or visit their Web site: www.mastergardeners.org.

16. McLean, *Medieval English Gardens,* p. 238; Madeleine Pelner Cosman, *Fabulous Feasts* (New York: George Braziller, 1976), pp. 48, 149; Culpeper, *Culpeper's Complete Herbal,* p. 176; Shakespeare, *Richard III,* III. 4, ll. 31–33, p. 556.

17. You can help maintain biodiversity by growing organic heirloom seeds. For more information, catalogs, and to order your own heirloom seeds, contact: Seeds of Change (888) 762-7333, www.seedsofchange.com; Garden City Seeds (406) 961-4837, www.gardencityseeds.com/home.htm; or Bountiful Gardens, 18001 Shafer Ranch Road, Willits, CA 95490-9626, (707) 459-6410, www.bountifulgardens.org.

18. For a beautiful personal response to the Rule of St.Benedict, see Kathleen Norris, *The Cloister Walk* (New York: Riverhead Books, 1996).

## Quotations

Milton, "Song on May Morning," ll. 5–6, p. 42.

Chaucer, Prologue to *Legend of Good Women,* ll. 36–39. A modern paraphrase would be: "In the joly time of May, when I hear the small birds sing, and see the flowers begin to spring up, farewell to my study for the rest of the season!"

Herbert, "The Flower," ll. 1–7, p. 165.

Herrick, "Corinna's Going a Maying," ll. 32–42, p. 68.

Chaucer, *Legend of Good Women,* ll. 180–84, p. 486. A modern paraphrase would be: "The long day I long for nothing else, and I shall not lie, but to look upon the daisy, that well by reason men call it the 'day's eye,' or else 'the eye of day.' "

Tusser, "The Good Huswifelie Physicke," p. 179.

Ben Jonson, "Song, to Celia," ll. 9–16 in *Seventeenth-Century Prose and Poetry,* p. 764.

Herbert, "The Rose," from *The Temple,* p. 178.

Tusser, *Five Hundred Points,* p. 106.

Ibid., p. 103.

Shakespeare, *I Henry IV,* II.5, ll. 365–66, p. 1186.

## Chapter 6: June

1. Spicer, *Yearbook,* pp. 241–43, 258; Urlin, *Festivals, Holy Days,* p. xiv.

2. Information about St. John's wort from "Herb News," *Prevention* (July 1999), p. 108; Dobelis, et al., *Magic and Medicine,* p. 20; and Lorna J. Sass, *To the King's Taste* (New York: Metropolitan Museum of Art, 1975), p. 131.

3. See Hildegard of Bingen, *Holistic Healing* (*Causae et Curae* [c. 1158]), ed. Mary Palmquist and John Kulas, O.S.B., trans. Manfred Pawlik, Patrick Madigan, S. J., and John Kulas, O.S.B. (Collegeville, Minn.: Liturgical Press, 1994), p. xviii, and Gabriele Uhlein, *Meditations with Hildegard of Bingen* (Santa Fe, N.M.: Bear, 1983), p. 17.

4. John Philpot Curran, *Speech upon the Right of Election of the Lord Mayor of Dublin* (1790), quoted in *Bartlett's Familiar Quotations*, 16th ed. ed. Justin Kaplan (Boston: Little, Brown, 1992), p. 351.

5. Sluggo snail bait is manufactured in Germany. It breaks down into iron phosphate in the garden and is nontoxic to pets. It should be available at your local garden center or contact Lawn and Garden Products, Inc., P.O. Box 35000, Fresno, CA 93745, (559) 499–2100.

6. McLean, *Medieval English Gardens*, p. 184; Rohde, *Herbs and Herb Gardening*, p. 123.

7. McLean, *Medieval English Gardens* p. 177; Peplow and Peplow, *In a Monastery Garden*, p. 110; Rohde, *Herbs and Herb Gardening*, p. 46.

8. McLean, *Medieval English Gardens*, p. 185; Peplow and Peplow, *In a Monastery Garden*, p. 110; Rohde, *Herbs and Herb Gardening*, p. 62; Dobelis et al., *Magic and Medicine*, p. 321.

Quotations

Herrick, "To the Virgins, to make much of Time," ll. 1–8, p. 84.

Chaucer, *The Parliament of Fowls*, ll. 183–86, p. 312. A modern paraphrase would be: "I saw a garden full of blossoming boughs beside a river in a green meadow. There was sweetness profound with flowers of white, blue, yellow, and red."

Dryden, *Virgil's Georgics*, II, ll. 49–50, p. 184.

Shakespeare, *Hamlet*, I.2, ll. 135–37, p. 1676.

*Chapter 7: July*

1. Eco, *Art and Beauty*, p. 46.

2. Bahouth, *Flowers, Birds, and Unicorns*, p. 69; McLean, *Medieval English Gardens*, p. 235; Sherrilyn Kenyon, *The Writer's Guide to Everyday Life in the Middle Ages* (Cincinnati, Ohio: Writer's Digest Books, 1995), p. 14; Arthur S. Way, trans., *The Science of Dining (Mensa Philosophica): A Medieval Treatise on the Hygiene of the Table and the Laws of Health* (London: Macmillan, 1936), p. 33; Culpeper, *Culpeper's Complete Herbal*, p. 252; Montagu Don, *The Sensuous Garden* (New York: Simon & Schuster, 1997), p. 52.

3. McLean, *Medieval English Gardens*, p. 40; Kenyon, *Writer's Guide*, p. 63.

4. The headquarters of Sunset Publications is 80 Willow Road, Menlo Park, CA 94025, (650) 321-3600. *Sunset's* gardens are open to the public Monday through Friday from 9:00 A.M. to 4:30 P.M. Admission is free. For information on gardens you can visit in your area, order a directory from The Garden Conservancy, P.O. Box 219, Cold Spring, NY 10516, (800) 842-2442. The directory is $12.95 plus $3.40 for shipping and handling.

5. Milton, *Paradise Lost*, IX, ll. 208–12, p. 383.

6. William Gouge, *Of Domestical Duties* (London: Haviland, 1622), pp. 88, 89. Cited in Diane Elizabeth Dreher, "Milton's Warning to Puritans in *Paradise Lost*: Another Look at the Separation Scene." *Christianity and Literature*, 41, no. 1 (Autumn 1991), p. 36.

7. Rohde, *Herbs and Herb Gardening*, pp. 34, 36, 43; McLean, *Medieval English Gardens*, p. 184; Peplow and Peplow, *In a Monastery Garden*, p. 110.

8. Way, *Science of Dining*, p. 29; McLean, *Medieval English Gardens*, p. 189.

9. Peplow and Peplow, *In a Monastery Garden*, p. 141; Landsberg, *Medieval Garden*, pp. 42–43.

10. McLean, *Medieval English Gardens*, p. 186; Rohde, *Herbs and Herb Gardening*, p. 102.

11. Sluggo snail bait, manufactured in Germany, is available in many garden centers in the United States. It's distributed by Lawn and Garden Products, Inc., P.O. Box 35000, Fresno, CA 93745, (559) 499-2100.

12. Some of the information about mulch is from Barbara Damrosch, *The Garden Primer* (New York: Workman, 1988), pp. 63–65, and Anne Halpin, *The Garden Year Planner* (New York: Putnam Berkeley, 1995), p. 30. These are both excellent references for general gardening information.

Quotations

Shakespeare, *The Winter's Tale*, IV.3. ll. 103–7, p. 2923.

Herrick, "The Succession of the foure sweet months," p. 23.

Ralph Austen, *A Treatise of Fruit Trees*, pt. 2, sig. f, quoted in Maren-Sofie Røstvig, *The Happy Man: Studies in the Metamorphosis of a Classical Ideal, 1600–1700* (Oslo: Akademisk forlag; Oxford: Blackwell, 1954), vol. 1, p. 183.

Dryden, *Virgil's Georgics*, I, ll. 157–62, p. 160.

Shakespeare, sonnet 94, ll. 9–14, pp. 1954–55.

*Chapter 8: August*

1. Thirteenth-century Book of Hours illustrated in Landsberg, *Medieval Garden*, pp. 94–96. See also Barbara Hanawalt, *The Ties That Bind: Peasant Families in Medieval England* (New York: Oxford Univ. Press, 1986), p. 126; Collins and Davis, *Medieval Book of Seasons*, p. 28; and McLean, *Medieval English Gardens*, p. 234.

2. Peplow and Peplow, *In a Monastery Garden*, pp. 60–61; McLean, *Medieval English Gardens*, p. 26; Umberto Eco, "How the Bean Saved Civilization," trans. William Weaver. *New York Times Magazine*, sec. 6 (April 18, 1999), pp. 136–38.

3. The *Sunset Western Garden Book* is a wonderful resource for gardeners in the American West, from the Pacific coast to the Rocky Mountain states, filled with color photographs for identifying plants, gardening advice, and a comprehensive encyclopedia of western plants. In 1997 Sunset published its *National Garden Book*, extending its coverage to the U.S. Midwest, South, Northeast, and Canada.

4. For information on beginning a walking program, see Alice Trevor, "Walk Off 10, 25, even 100 Pounds!" *Prevention* (October 1998), pp. 112–21, 176, or check out their Web site: www.healthyideas.com. Information on endorphin levels from Robert Numan, Ph.D., neuroscientist and marathon runner.

5. These are only a few versions of the circle of life. You probably know more. For the reference to *The Song of Lawino*, a book by the Ugandan poet Okot P'bitek (Nairobi, Kenya: East Africa Publishing House, 1966), I am grateful to my friend Elizabeth Moran, who used to teach African literature at Santa Clara University.

6. Ed Kleinschmidt Mayes's book *Works and Days* (Pittsburgh, Pa: Univ. of Pittsburgh Press, 1999) won the 1998 Associated Writing Program's Award in poetry.

7. McLean, *Medieval English Gardens*, p. 179.

Quotations

Herrick, "The Hock-Cart, or Harvest Home," ll. 1–6, p.101.

Tusser, *Five Hundred Points*, p. 124.

Herrick, "The Argument of His Book," ll. 1–4, from *Hesperides*, p. 5.

John Donne, "A Valediction Forbidding Mourning," ll. 35–36, from *Songs and Sonnets*, in *Seventeenth-Century Prose and Poetry*, p. 747.

Tusser, *Five Hundred Points*, pp. 124–25.

Herbert, "The Flower," ll. 43–49, p. 167.

Donne, "The Sunne Rising," ll. 9–10, p. 738.

*Chapter 9: September*

1. English proverb about St. Matthew's Day from Spicer, *Yearbook*, p. 134. Traditional harvest fairs included St. Giles's Fair at Oxford, and Widecombe Fair in Devonshire, as well as Sturbridge Fair in Cambridgeshire and Bridgewater Fair in Somerset, which dated back to the 1200s. For information about harvest fairs and other seasonal rituals, see Ibid., pp. 119, 127–28, 134–36; Urlin, *Festivals, Holy Days*, pp. 180–82.

2. Spicer, *Yearbook*, p. 257; M. McLeod Banks, *British Calendar Customs* (London: William Glaisher, 1937), vol.1, p. 55; Collins and Davis, *Medieval Book of Seasons*, pp. 26–28.

3. The Elizabeth F. Gamble Garden Center in Palo Alto is an urban estate donated to the public that has become a community center for gardening information and classes, as well as teas and special events. Don Ellis, the resident horticulturist, is happy to share his gardening expertise. Admission to the gardens is free, and they're open daily during daylight hours at 1431 Waverley Street, Palo Alto, CA 94301, (650) 329–1356. For information about botanical gardens, conservatories, and garden centers in your area, contact the Garden Conservancy, P.O. Box 219, Cold Spring, NY 10516, (914) 265–2029, fax (914) 265–9620.

4. For information on phototropism and heliotropism, see Raven et el., *Biology of Plants*, pp. 702–4, 722–23.

5. Martin E. P. Seligman, *Learned Optimism* (New York: Alfred A. Knopf, 1998), p. v. This is a wonderful guide to making a positive difference in your life.

6. Shakespeare, *The Winter's Tale*, IV.2. ll. 75, p. 2923.

7. Shakespeare, *Hamlet*, IV.5, l. 173, p. 1734.

8. Gerard, *Herball*, p. 1294.

9. Rohde, *Herbs and Herb Gardening*, pp. 13–20; 8, 26; McLean, *Medieval English Gardens*, p.194–95.

10. Gerard, *Herball*, p. 51, 59.

11. Quote and advice about dividing bulbs from Don Ellis, resident horticulturist at the Elizabeth F. Gamble Garden Center in Palo Alto, California. Recorded at a class on perennial maintenance on September 28, 1999, and used with permission.

12. For this recipe I am grateful to my longtime friend and colleague Professor Ann Brady of the English Department at Santa Clara University.

13. Gerard, *Herball*, p. 346.

14. Sass, *To the King's Taste*, pp. 106–7; McLean, *Medieval English Gardens*, pp. 226, 228; Gerard, *Herball*, p. 1460.

15. Peter Chan is an experimental biology technician at Portland State University. His book, *Peter Chan's Magical Landscape* (Pownall, Vt.: Storey Communications, 1988), tells how he transformed his unpromising yard into a paradise.

Quotations

Tusser, *Five Hundred Points*, p. 29.

Ibid., p. 35.

Ibid., p. 99.

*Chapter 10: October*

1. Spicer, *Yearbook*, pp. 153–54; Urlin, *Festivals, Holy Days*, p. 190.

2. Spicer, *Yearbook*, pp. 144–46, 152; Urlin, *Festivals, Holy Days*, pp. 187–88.

3. Way, *Science of Dining*, p. 28.

4. Dave is a pseudonym but his story is true, drawn from "Instant Interiors," a discussion of life in the Silicon Valley by Jane Meredith Adams, *San Francisco Focus* (November 1999), pp. 43–48.

5. Way, *Science of Dining*, pp. 26, 30.

6. Ibid., p. 25; Peplow and Peplow, *In a Monastery Garden*, p. 110.

7. Philippa Pullar, *Consuming Passions: A History of English Food and Appetite* (London: Hamish Hamilton, 1970), pp. 83–84; Collins and Davis, *Medieval Book of Seasons*, p. 106.

Quotations

Shakespeare, *Antony and Cleopatra*, V.2, ll. 85–87, p. 2699.

Herrick, "Upon Julia's Recovery," ll. 5–6, p. 7.

Herrick, "To Blossoms," ll. 1–6, p. 176.

Donne, "Elegy IX: The Autumnal," ll. 1–2, p. 752.

Herrick, "No Paines, no Gaines," p. 253.

Herrick, "Good precepts, or counsell," ll. 5–8, p. 247.

Herrick, "Out of Time, out of Tune," ll. 1–6, p. 279.

Tusser, *Five Hundred Points*, p. 40.

Chapter 11: *November*

1. My thanks to William Eisinger, Ph.D., professor of biology at Santa Clara University, for his assistance with this explanation.

2. Spicer, *Yearbook*, pp. 158–59.

3. Ibid., pp. 158, 164–65, 263, 269; Urlin, *Festivals, Holy Days*, pp. 205, 210, 216–17.

4. Sass, *To the King's Taste*, pp. 11, 19–20, 29–32; McLean, *Medieval English Gardens*, p. 200.

5. Menu for Richard II's feast on September 23, 1387, recorded in a medieval manuscript, *The Forme of Cury* (c. 1390), cited in Sass, *To the King's Taste*, pp. 19–20. Sass's book provides a wealth of information about medieval cuisine, including recipes we can use today.

6. Raven et al., *Biology of Plants*, p. 727.

7. For information on soil testing, organic amendments, and soil testing kits, contact Bountiful Gardens, 18001 Shafer Ranch Road, Willits, CA 95490, (707) 459–6410, www.bountifulgardens.org.

8. Raven et al., *Biology of Plants*, pp. 727, 736–38.

9. Ibid., p. 738.

10. Information on carbon-nitrogen ratios from K. A. Barbarick, "Organic Materials as Nitrogen Fertilizers," no. 0.546. Colorado State University Cooperative Extension. November 12, 1999, http://www.colostate.edu/Depts/CoopExt/, November 29, 1999.

11. The Santa Clara County Master Gardener Hotline is a service of the University of California Cooperative Extension Office at 1005 Timothy Drive, San Jose, CA 95133, available from 9:30 A.M. to 12:30 P.M., Pacific time, Monday through Friday at (408) 299-2638. Gardening information is also available on their Web site at www.mastergardeners.org.

12. Tusser, *Five Hundred Points*, p. 51; Thomas Jefferson, letter to Martha Jefferson Randolph, Philadelphia, July 21, 1793. Quoted by permission of the Massachusetts Historical Society.

13. Landsberg, *Medieval Garden*, pp. 96–97.

14. I got the idea for leaf mold from Pippa Greenwood, *Gardening Hints and Tips* (New York: DK Publishing, 1996), p. 42. This is an excellent source for all kinds of gardening advice.

15. Dryden, *Virgil's Georgics*, II, ll. 749–58, p. 206.

16. C. S. Lewis, *The Allegory of Love: A Study in Medieval Tradition* (Oxford: Oxford Univ. Press, 1936), p. 1. Used by permission.

Quotations

Ecclesiastes 3:1 in *King James Bible* (1611).

Herrick, "Content, not cates," ll. 1–2 p. 124.

Herrick, "Why Flowers change colour," ll. 1–4, p. 15.

Marvell, "Upon Appleton House," ll. 561–68, p. 972.

Herrick, "Rest Refreshes," ll. 1–2, p. 292.

Herbert, "Peace," ll. 37–42, p. 125.

Dryden, *Virgil's Georgics*, I, ll. 116–17, p. 158.

Ibid., I, ll. 118–19, p. 159.

*Chapter 12: December*

1. Shakespeare, sonnet 73, ll. 2–4, p. 1947.

2. Saturnalia from Urlin, *Festivals, Holy Days*, pp. 232–38.

3. St. Thomas Day rhyme from Urlin, *Festivals, Holy Days*, p. 225. For more about the Scandinavian origins of these later Christian customs, see Ibid., pp. 219–26, 232–38.

4. Spicer, *Yearbook*, p. 171–73; Urlin, *Festivals, Holy Days*, pp. 232–34, 243–46.

5. Pullar, *Consuming Passions*, p. 118; Spicer, *Yearbook*, pp. 184, 270; Urlin, *Festivals, Holy Days*, pp. 249, 256–57; Hassel, *Renaissance Drama*, p. 1.

6. For *Aloe vera* information, see John 19:39 in the Bible; Dobelis et al., *Magic and Medicine*, p. 83; Gerard, *Herball*, pp. 508–9; www.aloe-vera.org. Of course, for any serious ailment, you should see a doctor.

7. Tusser, *Five Hundred Points*, pp. 55–57.

Quotations

Shakespeare, *Love's Labour's Lost*, I.1, ll. 105–7, p. 743.

Dryden, *Virgil's Georgics*, I. ll. 403–6, p. 169.

Tusser, *Five Hundred Points*, p. 24.

Herrick, "A New-yeares gift sent to Sir Simeon Steward," ll. 45–50, p. 127.

Herrick, "Another," p. 264.

Herbert, "The Flower," ll. 8–14, p. 166.

Shakespeare, sonnet 97, ll. 1–4, p. 1955.

Milton, "Psalm I Done Into Verse," ll. 7–10, p. 162.

Dryden, *Virgil's Georgics*, II, ll. 280–81, 284–85, p. 190.

Herbert, "The Forerunners," ll. 34–36, p. 177.

Milton, "On the Morning of Christ's Nativity," ll. 61–63, p. 44.

Herbert, "An Offering," ll. 19–24, p. 147.


*Chapter 13: January*

1. Hassel, *Renaissance Drama*, p. 44; Spicer, *Yearbook*, pp. 9, 20, 266; Tusser, *Five Hundred Points*, p. 69.

2. Shakespeare, *I Henry IV*, I.2, ll. 182–83, p. 1164.

3. Spicer, *Yearbook*, pp. 24, 26.

4. Milton, *Paradise Lost*, XII, ll. 566–67, p. 467; Lao Tzu, *Tao Te Ching*, ch. 64, cited in Diane Dreher, *The Tao of Personal Leadership* (New York: HarperCollins, 1996), p. 26. For Bob's example, I am grateful to my husband, Dr. Robert Numan.

5. David S. Landes, *Clocks and the Making of the Modern World* (Cambridge, Mass.: Harvard Univ. Press, 1983), pp. 61, 68–69, 87, 89. Monastic life revolved around the canonical hours, eight periods of communal prayer. Specific times depended on the season, but the monks rose at 2:00 to 3:00 A.M. for vigils, said lauds at first light, prime at dawn, then terce around 9:00 A.M., sext at noon, and nones around 3:00, followed by vespers, and compline at dusk, after which they went to bed. The days also contained set times for work and holy reading. For more information about monastic life, see C. H. Lawrence, *Medieval Monasticism* (New York: Longmans, 1984), especially pp. 29–30, 98–99. For an excellent commentary on the Benedictine Rule, see Joan Chittister, *The Rule of Benedict: Insights for the Ages* (New York: Crossroad, 1992).

6. Landes, *Clocks*, pp. 67, 89, 114, 129.

7. Shakespeare, *Richard II*, III.4, ll. 35–40, 56–58, p. 991.

8. Herbert, "Paradise," pp. 132–33.

9. Seligman, *Learned Optimism*, pp. 82–87.

10. Information on medieval flower dyes from McLean, *Medieval English Gardens*, p. 35.

11. For information on Newton and the electromagnetic spectrum, see R. L. Gregory, *Eye and Brain: The Psychology of Seeing*, 3d ed. (New York: McGraw-Hill, 1978), pp. 15–25.

12. Derek Fell's *Essential Gardener* (New York: Cresent Books, 1990) contains color lists for annuals, bulbs, and perennials as well as excellent diagrams of garden designs.

13. Tusser, *Five Hundred Points*, p. 179.

14. For more ideas for soups with recipes for all seasons, see Victor-Antoine d'Avila-Latourrette, *Twelve Months of Monastery Soups* (New York: Broadway Books, 1998).

Quotations

Shakespeare, sonnet 5, ll. 5–8, p. 1925.

Herrick, "A New-yeare's gift sent to Sir Simeon Steward," ll. 17–26, p. 127.

Herrick, "St. Distaffs Day, or the morrow after Twelfth day," ll. 11–14, p. 315.

Herrick, "The Frozen Heart," ll. 1–4, p. 8.

Tusser, *Five Hundred Points*, p. 177.

Dryden, *Virgil's Georgics*, II, ll. 85–86, p. 183.

Ibid., ll. 502–10, p. 197.

*Chapter 14: February*

1. Henry David Thoreau, "Solitude," in *Walden*, ed. Sherman Paul (Boston: Houghton Mifflin, 1960), p. 91.

2. Spicer, *Yearbook*, pp. 30–31, 258.

3. Ibid., p. 33.

4. Ibid., p. 37.

5. Hyams, *History of Gardens*, p. 4.

6. Ibid., p. 5.

7. Carts with fluorescent growth lights, other indoor growing supplies, and helpful advice are available from Shepherd Seeds, 30 Irene Street, Torrington, CT 06790-6658, (860) 482-3638, www.shepherdseeds.com, and Park Seed, 1 Parkton Avenue, Greenwood, SC 29647-0001, (800) 845-3369, www.parkseed.com.

8. A wonderful resource for starting seeds indoors is Nancy Bubel, *The New Seed Starter's Handbook* (Emmaus, Pa.: Rodale Press, 1988).

9. Nelson Mandela, *Long Walk to Freedom* (Boston: Little, Brown, and Company 1994), pp. 425–26, 449. Used by permission.

10. Hanawalt, *Ties That Bind*, p. 126; Landsberg, *Medieval Garden*, p. 96.

11. Tusser, *Five Hundred Points*, p. 80.

12. McLean, *Medieval English Gardens*, p. 245.

13. The Plexiglas sprouter and seeds for broccoli, radish, mung bean, watercress, and other sprouts are available from Wayside Gardens, 1 Garden Lane, Hodges, SC 29695-0001, (800) 845-1124, www.waysidegardens.com.

Quotations

Herrick, "Hope well and Have well: or, Faire after Foule weather," p. 188.

Milton, "Il Penseroso," ll. 126–30, p. 75.

Shakespeare, *Much Ado About Nothing*, V.4, ll. 40–42, p.1441.

Herrick, "Ceremonies for Candlemasse Eve," ll. 1–8, 21–22, p. 285.

Herrick, "To his Valentine, on S. Valentine's Day," ll. 1–4, p. 149.

Shakespeare, *A Midsummer Night's Dream*, IV.1, ll. 136–37, p. 848.

Ibid., II.1, ll. 249–50, p. 827.

Donne, "The Ecstasy," ll. 37–40, p. 748.

Milton, *Paradise Lost*, IV, ll. 214–18, p. 283.

Herbert, "The Flower," ll. 36–42, p. 166.

# USDA Plant Hardiness Zone Map

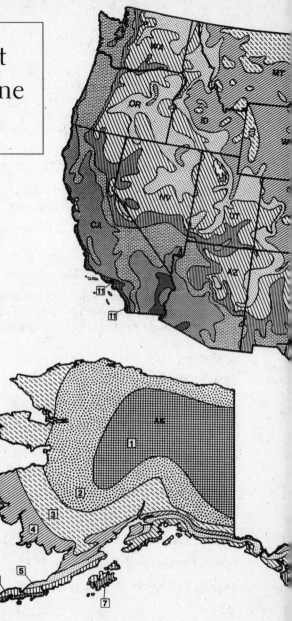

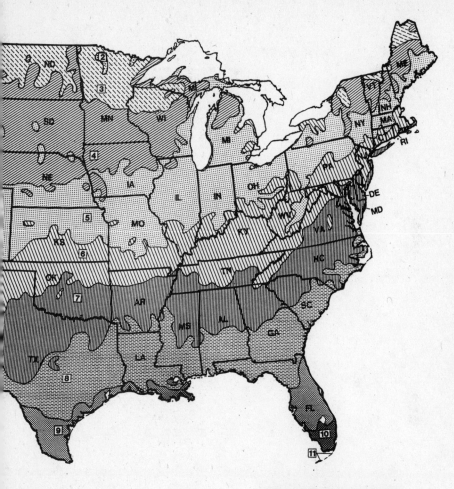

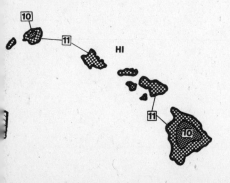

RANGE OF AVERAGE ANNUAL MINIMUM
TEMPERATURES FOR EACH ZONE

| ZONE 1 | BELOW -50°F | |
| --- | --- | --- |
| ZONE 2 | -50° TO -40° | |
| ZONE 3 | -40° TO -30° | |
| ZONE 4 | -30° TO -20° | |
| ZONE 5 | -20° TO -10° | |
| ZONE 6 | -10° TO 0° | |
| ZONE 7 | 0° TO 10° | |
| ZONE 8 | 10° TO 20° | |
| ZONE 9 | 20° TO 30° | |
| ZONE 10 | 30° TO 40° | |
| ZONE 11 | ABOVE 40° | |

$\mathscr{P}ermissions$

The author and publisher gratefully acknowledge permission to use selections and information from the following copyrighted materials:

Page 15: John Dryden, poetry from *Virgil's Georgics (1697)* in *The Works of John Dryden*, ed. H. T. Swedenberg et al. Vol V (Berkeley and Los Angeles: Univ. of California Press, 1987). Used by permission of the Regents of the University of California and the University of California Press.

Page 32: St. Bonaventura, "Sermon on St. Francis, October 4, 1255" from *The Disciple and the Master: St. Bonaventura's Sermons on St. Francis of Assisi*. Trans. and ed. Eric Doyle, O.F.M. (Chicago: Franciscan Herald Press, 1983), p. 62. Included by permission of the Franciscan Press, Quincy University, Quincy, Ill. 62301.

Page 34: Thomas Tusser, selections from *Five Hundred Points of Good Husbandry (1580)*. Introd. Geoffrey Grigson (Oxford: Oxford University Press, © 1984). Used by permission of Oxford University Press.

Page 35: John Jeavons, description of double-digging from *How to Grow More Vegetables Than You Ever Thought Possible on Less Land Than You Can Imagine* (Berkeley, Calif.: Ten Speed Press, © 1995 Ecology Action), pp. 7–14. Used by permission of John Jeavons, Ecology Action, and Ten Speed Press.

Page 44: Geoffrey Chaucer, selections from *The Works of Geoffrey Chaucer*, ed. F. N. Robinson. Copyright © 1957 by Houghton Mifflin Company. Used with permission.

Page 45: John Milton, selections from *John Milton: Complete Poems and Major Prose*. ed. Merritt Y. Hughes (New York: Odyssey Press, 1957; Prentice-Hall, © 2000). Reprinted by permission of Prentice-Hall, Inc., Upper Saddle River, N.J.

# Index

# Index